Cambridge Tr
and Math

GENERAL EDITORS
F. SMITHIES PH.D AND J. A. TODD, F.R.S.

No. 41

INTRODUCTION TO MODERN PRIME NUMBER THEORY

INTRODUCTION TO MODERN PRIME NUMBER THEORY

BY

T. ESTERMANN, D.Sc.
Professor of Mathematics in the University of London

CAMBRIDGE
AT THE UNIVERSITY PRESS
1969

CAMBRIDGE UNIVERSITY PRESS
Cambridge, New York, Melbourne, Madrid, Cape Town, Singapore,
São Paulo, Delhi, Dubai, Tokyo, Mexico City

Cambridge University Press
The Edinburgh Building, Cambridge CB2 8RU, UK

Published in the United States of America by Cambridge University Press, New York

www.cambridge.org
Information on this title: www.cambridge.org/9780521168281

First published 1952
First paperback edition 2010

A catalogue record for this publication is available from the British Library

ISBN 978-0-521-07735-4 Hardback
ISBN 978-0-521-16828-1 Paperback

PREFACE TO THE SECOND IMPRESSION

§§ 2·5–2·7 have been re-written. Otherwise there have been only a few minor changes.

PREFACE

This book takes the reader as far as Vinogradoff's theorem, that every sufficiently large odd positive integer can be represented as a sum of three primes. It assumes nothing of the theory of numbers that is not given in Hardy and Wright, *An Introduction to the Theory of Numbers* (Oxford, 1938), hereafter quoted as H.-W.

My main purpose in writing this book was to enable those mathematicians who are not specialists in the theory of numbers to learn some of its non-elementary results and methods without too great an effort.

Chapter 1 deals with a refinement of the prime number theorem, obtained by de la Vallée Poussin in 1899 (*Mém. cour. Acad. R. Belg.* **59**, 1), three years after the discovery by him and Hadamard of the prime number theorem itself. The method of proof used here is essentially due to Landau.

The main result of Chapter 2, Theorem 55, with its emphasis on uniformity in k, was stated and proved in 1936 by Walfisz (*Math. Z.* **40**, 598, Hilfssatz 3), but the difficulty in obtaining it had then been removed by Siegel (*Acta Arith.* **1** (1935), 83–6), who had discovered a property of Dirichlet's L functions which led at once to Theorem 48. The method used in §§ 2·5–2·6 is taken from the theory of groups. This theory is not assumed, but the reader who finds §§2·5–2·7 difficult may be referred to an alternative proof of Theorem 28: Landau, *Vorlesungen über Zahlentheorie* (Leipzig, 1927), **1**, Satz 134.

The method of Chapter 3, excluding Theorem 56, is due to Hardy and Littlewood (*Acta Math., Stockh.*, **44** (1923), 1–70). Vinogradoff, in obtaining his famous result (*Rec. Math.* T. 2 (44), 2 (1937), 179–95), built on foundations laid by them. At the same time Theorem 56, which is due to him, is a very substantial contribution.

I am indebted to my colleague Mr. H. Kestelman for valuable advice and criticism, and to Mr. R. C. Wellard for checking part of the manuscript and correcting a mistake.

T. ESTERMANN

London, August 1951

CONTENTS

REMARKS ON NOTATION

Throughout this book, the following letters denote the following types of number:

h, j, l, m, n = integers;
k, q = positive integers;
p = primes;
$t, u, v, x, y, \sigma, \theta$ = real numbers;
M, ϵ, δ = positive numbers;
s, w, z = complex numbers.

The real and imaginary parts of s are, as usual, denoted by σ and t respectively.

$C_1, C_2, \ldots$ are suitable (sufficiently large) positive absolute constants.

$\int_z^w$ denotes an integral taken along the straight line from z to w.

$[x]$ denotes the greatest integer less than or equal to x.

$e(z)$ is an abbreviation for $e^{2\pi iz}$.

INTRODUCTION TO MODERN PRIME NUMBER THEORY

CHAPTER 1

THE RIEMANN ZETA FUNCTION AND A REFINEMENT OF THE PRIME NUMBER THEOREM

1·1. The prime number theorem states that $\pi(m)$, the number of primes not exceeding m, is asymptotic to $m/\log m$. Our object is to obtain a better approximation to $\pi(m)$, and we shall show that

$$\pi(m) = \sum_{n=2}^{m} \frac{1}{\log n} + O(me^{-c\sqrt{\log m}}), \tag{1}$$

where c is a suitable positive constant. This is a refinement of the prime number theorem, for it is easily seen that

$$\sum_{n=2}^{m} \frac{1}{\log n} \sim \frac{m}{\log m}.$$

The numerical value of c is unimportant. In fact, (1) is true for any positive constant c, and this can be shown by a slight modification of the method used here. Still better results in this direction can be proved by more complicated methods.

The proof of (1) is mostly analytical. Only the last step is elementary, and consists of a straightforward argument which deduces (1) from the formula

$$\psi(m) = m + O(me^{-c\sqrt{\log m}}), \tag{2}$$

where

$$\psi(m) = \sum_{n=1}^{m} \Lambda(n), \tag{3}$$

$\Lambda(p^k) = \log p$, and $\Lambda(n) = 0$ if n is not of the form p^k. The analytical part of the proof depends on certain properties of the Riemann zeta function, originally defined by

$$\zeta(s) = \sum_{n=1}^{\infty} n^{-s} \quad (\sigma > 1). \tag{4}$$

We shall extend this definition by analytic continuation up to the imaginary axis. The further analytic continuation of the zeta function into the left-hand half-plane, though well known, will not be given in this book, as it is not needed for our purpose. The deepest property of the zeta function used here is that

$$\zeta(s) \neq 0 \quad \{\sigma > 1 - 1/(C_1 \log|t|),\ |t| > C_1\}. \tag{5}$$

1·2. The details are as follows. In order to obtain the analytic continuation of $\zeta(s)$, we consider the functions

$$f_n(s) = n^{-s} - \int_n^{n+1} u^{-s}\,du \quad (n = 1, 2, \ldots). \tag{6}$$

It is obvious that
$$\zeta(s) = \sum_{n=1}^{\infty} f_n(s) + \frac{1}{s-1} \tag{7}$$

if $\sigma > 1$. Also
$$f_n(s) = \int_n^{n+1} (n^{-s} - u^{-s})\,du$$

and

$$|n^{-s} - u^{-s}| = \left|\int_n^u s v^{-s-1}\,dv\right| \leqslant |s| \int_n^{n+1} v^{-\sigma-1}\,dv \quad (n \leqslant u \leqslant n+1),$$

so that
$$|f_n(s)| \leqslant |s| \int_n^{n+1} v^{-\sigma-1}\,dv. \tag{8}$$

We use the term '*locally uniformly at s_0*' for 'uniformly in some circle about s_0' and the term '*locally uniformly in S*' (where S is a set of points) for 'locally uniformly at all points of S'. It follows from (8) that $\sum_{n=1}^{\infty} f_n(s)$ converges locally uniformly in the half-plane $\sigma > 0$. Hence we may *define* $\zeta(s)$ for $0 < \sigma \leqslant 1$, $s \neq 1$, by (7), thus obtaining the desired analytic continuation and

THEOREM 1. *$\zeta(s)$ is regular for $\sigma > 0$, except at $s = 1$, where it has a simple pole with residue* 1.

We also deduce from (7) and (8) that

$$\left|\zeta(s) - \frac{1}{s-1}\right| \leqslant |s| \int_1^\infty v^{-\sigma-1}\,dv = \frac{|s|}{\sigma} \quad (\sigma > 0,\ s \neq 1). \tag{9}$$

The next theorem, which belongs to the elementary theory of numbers, will be used here repeatedly.

THEOREM 2. *If $f(n)$ is multiplicative, and $\sum_{n=1}^{\infty} |f(n)|$ converges, then*

$$\sum_{n=1}^{\infty} f(n) = \prod_p \sum_{m=0}^{\infty} f(p^m).$$

This follows from H.-W., Theorem 286, on putting $s = 0$ and noting that $f(1) = 1$ for any multiplicative function $f(n)$ which does not vanish identically. Conversely, Theorem 2 implies H.-W., Theorem 286, since n^{-s} is a multiplicative function of n.

THEOREM 3. *Let $\sigma > 1$. Then*

$$\zeta(s) = \prod_p (1 - p^{-s})^{-1}.$$

This follows from (4) and Theorem 2. We deduce that

$$\zeta(s) \neq 0 \quad (\sigma > 1). \tag{10}$$

THEOREM 4. *Let $\sigma > 1$. Then*

$$\frac{\zeta'(s)}{\zeta(s)} = -\sum_p \frac{\log p}{p^s - 1}.$$

This follows from Theorem 3 since the series converges locally uniformly in the half-plane $\sigma > 1$. We deduce that

$$\frac{\zeta'(s)}{\zeta(s)} = -\sum_{n=1}^{\infty} \Lambda(n) n^{-s} \quad (\sigma > 1) \tag{11}$$

(cf. H.-W., Theorem 294, where s is, however, restricted to real values).

1·3. The next theorem, which is almost trivial, will help us to establish two further properties of the zeta function needed in our proof of (5).

THEOREM 5. *Let* $|z| = 1$. *Then* $\mathbf{R}(3+4z+z^2) \geqslant 0$.

Proof. Putting $z = x+iy$, we have $x^2+y^2 = 1$ and hence

$$\mathbf{R}(3+4z+z^2) = 3+4x+x^2-y^2 = 2+4x+2x^2 = 2(1+x)^2 \geqslant 0.$$

The two properties referred to are as follows:

THEOREM 6. *Let* $u > 1$. *Then*

$$\mathbf{R}\left\{3\frac{\zeta'(u)}{\zeta(u)}+4\frac{\zeta'(u+iv)}{\zeta(u+iv)}+\frac{\zeta'(u+2iv)}{\zeta(u+2iv)}\right\} \leqslant 0 \tag{12}$$

and

$$|\,\zeta^3(u)\,\zeta^4(u+iv)\,\zeta(u+2iv)\,| \geqslant 1. \tag{13}$$

Proof. By (11),

$$3\frac{\zeta'(u)}{\zeta(u)}+4\frac{\zeta'(u+iv)}{\zeta(u+iv)}+\frac{\zeta'(u+2iv)}{\zeta(u+2iv)} = -\sum_{n=1}^{\infty}\Lambda(n)\,n^{-u}a_n,$$

where $a_n = 3+4n^{-iv}+n^{-2iv}$, so that, by Theorem 5, $\mathbf{R}a_n \geqslant 0$. This proves (12). Also, by Theorem 3,

$$\zeta(s) = \exp\sum_p \log(1-p^{-s})^{-1} = \exp\sum_p\sum_{m=1}^{\infty}\frac{p^{-ms}}{m}$$

$$= \exp\sum_{n=2}^{\infty}\frac{\Lambda(n)}{\log n}n^{-s} \quad (\sigma > 1),$$

so that

$$|\,\zeta^3(u)\,\zeta^4(u+iv)\,\zeta(u+2iv)\,| = \exp\sum_{n=2}^{\infty}\frac{\Lambda(n)}{\log n}n^{-u}\mathbf{R}a_n$$

with a_n defined as before. Since $\mathbf{R}a_n \geqslant 0$, this proves (13).

1·4. The next four theorems belong to the theory of functions. The third will be the main analytical tool in the proof of (5). The fourth will only be used in the next chapter.

THEOREM 7. *Let* $r>1$,

$$f(z)=\sum_{n=1}^{\infty}b_n z^n \quad (|z|<r), \quad \textit{and} \quad \mathbf{R}f(z)\leqslant M \quad (|z|=1).$$

Then $$|b_n|\leqslant 2M \quad (n=1,2,\ldots).$$

Proof. Putting $b_n=|b_n|e^{i\theta_n}$, we have

$$\mathbf{R}f(e^{i\theta})=\sum_{n=1}^{\infty}|b_n|\cos(\theta_n+n\theta),$$

and this series converges uniformly, so that

$$\int_0^{2\pi}\mathbf{R}f(e^{i\theta})\,d\theta=0$$

and $$\int_0^{2\pi}\mathbf{R}f(e^{i\theta})\cos(\theta_n+n\theta)\,d\theta=\pi|b_n| \quad (n=1,2,\ldots),$$

which implies that

$$\pi|b_n|=\int_0^{2\pi}\mathbf{R}f(e^{i\theta})\{1+\cos(\theta_n+n\theta)\}\,d\theta$$
$$\leqslant\int_0^{2\pi}M\{1+\cos(\theta_n+n\theta)\}\,d\theta=2\pi M,$$

and the result follows.

THEOREM 8. *Let* $r>1$, *let* $g(z)$ *be regular for* $|z|<r$, *and let* $g(z)\neq 0$ $(|z|<r)$ *and* $|g(z)/g(0)|\leqslant e^M$ $(|z|=1)$. *Then* $|g'(0)/g(0)|\leqslant 2M$.

This follows from Theorem 7 with

$$f(z)=\int_0^z\frac{g'(w)}{g(w)}\,dw \quad \text{and} \quad b_n=f^{(n)}(0)/n!,$$

which implies that $$g'(0)/g(0)=b_1.$$

THEOREM 9. *Let* $f(z)$ *be regular, and* $|f(z)/f(0)|\leqslant e^M$, *for* $|z|\leqslant 2$; *let* $0<a\leqslant 1$,

$$f(z)\neq 0 \quad (|z|\leqslant 1,\ \mathbf{R}z>0), \tag{14}$$

and let $f(z)$ *have a zero of order* h *at* $z=-a$. *Then*

$$-\mathbf{R}\{f'(0)/f(0)\}\leqslant 2M-h/a.$$

Proof. Let the zeros of $f(z)$ within and on the circle $|z| = 1$ be $z_1, z_2, \ldots, z_l$, of orders $k_1, k_2, \ldots, k_l$ respectively, and put

$$g(z) = f(z) \prod_{m=1}^{l} (z - z_m)^{-k_m}.$$

Then $g(z)$ is regular for $|z| \leqslant 2$, and there is a number $r > 1$ such that $g(z) \neq 0 \quad (|z| < r)$. Also

$$\left|\frac{g(z)}{g(0)}\right| = \left|\frac{f(z)}{f(0)}\right| \prod_{m=1}^{l} \left(\frac{|z_m|}{|z - z_m|}\right)^{k_m} \leqslant \left|\frac{f(z)}{f(0)}\right| \leqslant e^M \quad (|z| = 2),$$

and hence, by the maximum modulus theorem,

$$|g(z)/g(0)| \leqslant e^M \quad (|z| = 1).$$

From this and Theorem 8 it follows that

$$-\mathbf{R}\{g'(0)/g(0)\} \leqslant |g'(0)/g(0)| \leqslant 2M,$$

which means that $$-\mathbf{R}\left\{\frac{f'(0)}{f(0)} + \sum_{m=1}^{l} \frac{k_m}{z_m}\right\} \leqslant 2M,$$

so that $$-\mathbf{R}\frac{f'(0)}{f(0)} \leqslant 2M + \sum_{m=1}^{l} k_m \mathbf{R}\frac{1}{z_m}.$$

Now, by (14), all the terms $k_m \mathbf{R}(1/z_m)$ are negative or 0, and $-h/a$ is one of them. Hence the result.

THEOREM 10. *Let $f(z)$ be regular, and $|f(z)/f(0)| \leqslant e^M$, for $|z| \leqslant 2$; let $|a| \leqslant 1$, $|b| \leqslant 1$, $a \neq b$, $f(a) = f(b) = 0$, and let (14) hold. Then $-\mathbf{R}\{f'(0)/f(0)\} \leqslant 2M + \mathbf{R}(1/a) + \mathbf{R}(1/b)$.*

This can be proved in the same way as Theorem 9.

1·5. We are now in a position to prove a theorem which implies (5). For convenience, we put

$$t^* = \max(|t|, 100) \tag{15}$$

and $$\eta(s) = (s-1)\zeta(s) \quad (s \neq 1), \quad \eta(1) = 1. \tag{16}$$

Theorem 11. *$\zeta(s)$ has no zeros in the set of points D given by $\sigma > 1 - 1/(4000 \log t^*)$.*

Proof. Suppose that the theorem is false. Then there are numbers σ_0 and t_0 such that

$$\zeta(\sigma_0 + it_0) = 0 \tag{17}$$

and
$$\sigma_0 > 1 - 1/(4000 \log t_0^*), \tag{18}$$

where
$$t_0^* = \max(|t_0|, 100). \tag{19}$$

It follows from (10) and (17) that

$$\sigma_0 \leqslant 1, \tag{20}$$

and from this and (9), (17), and (18) that $|t_0| \geqslant \frac{3}{4}$. Hence, if $1 < u < 2$ and $|s - u - it_0| \leqslant \frac{1}{4}$, then $|s-1| \geqslant |t| \geqslant |t_0| - \frac{1}{4} \geqslant \frac{1}{2}$, $|s| \leqslant 3 + t_0^*$, and $\sigma > \frac{3}{4}$, so that, by (9) and (19),

$$|\zeta(s)| \leqslant 6 + \tfrac{4}{3}t_0^* \leqslant \tfrac{3}{2}t_0^* \quad (1 < u < 2,\ |s-u-it_0| \leqslant \tfrac{1}{4}). \tag{21}$$

Similarly
$$|\zeta(s)| \leqslant 3t_0^* \quad (1 < u < 2,\ |s-u-2it_0| \leqslant \tfrac{1}{4}). \tag{22}$$

Also, if $1 < u$ and $|s-u| \leqslant \frac{1}{4}$, then

$$\frac{\sigma^2}{|s|^2} \geqslant \frac{\sigma^2}{|s|^2} - \left(\frac{\sigma}{|s|} - \frac{|s|}{u}\right)^2 = 1 - \frac{|s-u|^2}{u^2} > \frac{15}{16}$$

($\frac{15}{16}$ is the square of the cosine of half the angle subtended at the origin by the circle $|s-1| = \frac{1}{4}$), so that

$$|s|/\sigma < \sqrt{(16/15)} < 31/30.$$

Hence, by (16) and (9),

$$|\eta(s)| \leqslant 1 + \tfrac{31}{30}|s-1| < \tfrac{4}{3} \quad (1 < u < \tfrac{21}{20},\ |s-u| \leqslant \tfrac{1}{4}). \tag{23}$$

Now let
$$u = 1 + 1/(800 \log t_0^*) \tag{24}$$

and
$$f(z) = \eta^3(u + \tfrac{1}{8}z)\, \zeta^4(u + it_0 + \tfrac{1}{8}z)\, \zeta(u + 2it_0 + \tfrac{1}{8}z). \tag{25}$$

2

Then, by (23), (21), (22), and (19),

$$|f(z)| \leqslant (\tfrac{4}{3})^3 (\tfrac{3}{2} t_0^*)^4 3t_0^* < t_0^{*6} \quad (|z| \leqslant 2). \tag{26}$$

Also, putting

$$g(w) = \zeta^3(u+w)\,\zeta^4(u+it_0+w)\,\zeta(u+2it_0+w), \tag{27}$$

we have, by (25) and (16),

$$f(z) = (u-1+\tfrac{1}{8}z)^3 g(\tfrac{1}{8}z),$$

which implies that $$f(0) = (u-1)^3 g(0) \tag{28}$$

and $$\frac{f'(0)}{f(0)} = \frac{3}{8(u-1)} + \frac{1}{8}\frac{g'(0)}{g(0)}. \tag{29}$$

Now, by (27) and Theorem 6, $|g(0)| \geqslant 1$ and $\mathbf{R}\{g'(0)/g(0)\} \leqslant 0$. Hence, by (24), (28), and (29),

$$|f(0)| \geqslant (800 \log t_0^*)^{-3} \tag{30}$$

and $$\mathbf{R}\{f'(0)/f(0)\} \leqslant 300 \log t_0^*. \tag{31}$$

Also $x/\log x$ increases with x when $x \geqslant e$. Hence

$$t_0^*/\log t_0^* \geqslant 100/\log 100 > 20,$$

and hence, by (30), $|f(0)| > (40t_0^*)^{-3} > t_0^{*-6}$. From this and (26) it follows that

$$|f(z)/f(0)| \leqslant t_0^{*12} \quad (|z| \leqslant 2). \tag{32}$$

Now, by (17) and (25), $f(z)$ has a zero at $z = 8(\sigma_0 - u)$, and, denoting the order of this zero by h, we have

$$h \geqslant 4. \tag{33}$$

Putting $M = 12 \log t_0^*$ and $a = 8(u-\sigma_0)$, we can now verify that all the hypotheses of Theorem 9 are satisfied. In fact, $f(z)$ is regular for $|z| \leqslant 2$ by (25), (16), and Theorem 1, since $u > 1$ and $|t_0| \geqslant \frac{3}{4}$; $|f(z)/f(0)| \leqslant e^M (|z| \leqslant 2)$ by (32); $0 < a \leqslant 1$ by (24), (20), (18), and (19); (14) follows from (25), (16), and (10) since $u > 1$, and we have just seen that $f(z)$ has a zero of order h at $z = -a$. Hence, by Theorem 9, (31), and (33),

$$0 \leqslant 2M - \frac{4}{a} + 300 \log t_0^* = 324 \log t_0^* - \frac{1}{2(u-\sigma_0)}.$$

From this and (24) and (18) it follows that

$$324 \geqslant \frac{1}{2(u-\sigma_0)\log t_0^*} > \frac{1}{2(800^{-1}+4000^{-1})} = \frac{1000}{3}.$$

This is a contradiction; so Theorem 11 is proved.

THEOREM 12. *$\eta'(s)/\eta(s)$ is regular in the set of points D of Theorem* 11.

This follows from (16) and Theorems 1 and 11.

1·6. We shall deduce that $\eta'(s)/\eta(s) = O(\log^3 t^*)$ in a suitable subset of D. First, however, we have to prove another theorem from the theory of functions.

THEOREM 13. *Let $0 < r_1 < r_2$, let $g(z)$ be regular for $|z| < r_2$ and let $g(0) = 0$ and $\mathbf{R}g(z) \leqslant M \quad (|z| = r_1)$. Then*

$$|g'(z)| \leqslant 2Mr_1(r_1 - |z|)^{-2} \quad (|z| < r_1).$$

Proof. Let $f(z) = g(r_1 z)$ and $r = r_2/r_1$. Then there are numbers $b_1, b_2, \ldots$ such that all the hypotheses of Theorem 7 are satisfied. Hence, if $|z| < r_1$, we have

$$|g'(z)| = \frac{1}{r_1}\left|f'\left(\frac{z}{r_1}\right)\right| = \frac{1}{r_1}\left|\sum_{n=1}^{\infty} nb_n\left(\frac{z}{r_1}\right)^{n-1}\right| \leqslant \frac{1}{r_1}\sum_{n=1}^{\infty} n\,|b_n|\left(\frac{|z|}{r_1}\right)^{n-1}$$

$$\leqslant \frac{2M}{r_1}\sum_{n=1}^{\infty} n\left(\frac{|z|}{r_1}\right)^{n-1} = 2Mr_1(r_1 - |z|)^{-2},$$

which proves the theorem.

THEOREM 14. *Let* $1-1/(10000\log t^*) \leqslant \sigma < 2$. *Then*

$$|\eta'(s)/\eta(s)| \leqslant C_2 \log^3 t^*.$$

Proof. Let $r_1 = 1+1/(5000\log t^*)$, $r_2 = 1+1/(4500\log t^*)$, and

$$g(z) = \int_{2+it}^{2+it+z} \frac{\eta'(w)}{\eta(w)}\,dw. \tag{34}$$

Then it is easily seen that any point w for which $|w-(2+it)| < r_2$ is a point of D (the set of points defined in Theorem 11). From this and Theorem 12 and (34) it follows that $g(z)$ is regular for $|z| < r_2$. Also, by (34), $e^{g(z)} = \eta(2+it+z)/\eta(2+it)$. Now, by (16), (9), and (15),

$$|\eta(2+it+z)| \leqslant 1+\frac{(4+|t|)^2}{2-r_1} < 2t^{*2} \quad (|z| \leqslant r_1),$$

and, by (16) and Theorem 3,

$$|\eta(2+it)|^{-1} \leqslant |\zeta(2+it)|^{-1}$$
$$\leqslant \prod_p (1+p^{-2}) < \prod_p (1-p^{-2})^{-1} = \zeta(2) < 2.$$

Hence $e^{\mathbf{R}g(z)} < 4t^{*2} < t^{*3}$ $\quad (|z| = r_1)$, and the hypotheses of Theorem 13 are satisfied with $M = 3\log t^*$. It follows that

$$|\eta'(s)/\eta(s)| = |g'(s-2-it)| = |g'(\sigma-2)|$$
$$\leqslant 2Mr_1(r_1-2+\sigma)^{-2} \leqslant 2Mr_1(10000\log t^*)^2 < 10^9 \log^3 t^*,$$

which proves the theorem.

1·7. Our next task is to prove (2). The connexion between $\zeta(s)$ and $\psi(m)$ may be expressed in the formula

$$\psi(m) = -\frac{1}{2\pi i}\int_{a-i\infty}^{a+i\infty} \frac{(m+\frac{1}{2})^s}{s}\frac{\zeta'(s)}{\zeta(s)}\,ds \quad (a>1).$$

We shall neither prove nor use this formula. Instead, we consider the integral

$$\int_{a-ib}^{a+ib} \frac{(m+\frac{1}{2})^s}{s}\frac{\zeta'(s)}{\zeta(s)}\,ds,$$

which, for suitable values of a and b, provides a good enough approximation to $-2\pi i\psi(m)$. Then we replace $\zeta'(s)/\zeta(s)$ by $-1/(s-1)+\eta'(s)/\eta(s)$, using (16), and deduce from Theorem 12 and Cauchy's theorem that

$$\int_{a-ib}^{a+ib}\frac{(m+\frac{1}{2})^s}{s}\frac{\eta'(s)}{\eta(s)}\,ds=\int_C\frac{(m+\frac{1}{2})^s}{s}\frac{\eta'(s)}{\eta(s)}\,ds,$$

where C is a suitable broken line. Finally we obtain an inequality for the last integral from Theorem 14. These are the main steps in the proof of (2).

For convenience, we put

$$E(x)=\begin{cases}1 & (x>1),\\ 0 & (0<x<1).\end{cases} \tag{35}$$

Then, by (3),

$$\psi(m)=\sum_{n=1}^{\infty}E\left(\frac{m+\frac{1}{2}}{n}\right)\Lambda(n)\quad (m=1,2,\ldots). \tag{36}$$

THEOREM 15. *Let $a>0$, $b>0$, $x>0$, and $x\neq 1$. Then*

$$\left|\int_{a-ib}^{a+ib}\frac{x^s}{s}\,ds-2\pi iE(x)\right|\leqslant\frac{2x^a}{b\,|\log x|}.$$

Proof. Suppose, first, that $x>1$. Then it easily follows from the theorem of residues that

$$J_1+\int_{a-ib}^{a+ib}\frac{x^s}{s}\,ds+J_2=2\pi i,$$

where $$J_1=\int_{-\infty-ib}^{a-ib}\frac{x^s}{s}\,ds,\quad J_2=\int_{a+ib}^{-\infty+ib}\frac{x^s}{s}\,ds.$$

Now $$|J_1|=\left|\int_{-\infty}^{a}\frac{x^{\sigma-ib}}{\sigma-ib}\,d\sigma\right|\leqslant\int_{-\infty}^{a}\frac{x^\sigma}{b}\,d\sigma=\frac{x^a}{b\log x}.$$

Similarly, $|J_2|\leqslant x^a/(b\log x)$, and the result follows in this case.

Now suppose that $0<x<1$. Then it easily follows from Cauchy's theorem that

$$J_3+\int_{a-ib}^{a+ib}\frac{x^s}{s}\,ds+J_4=0,$$

where
$$J_3=\int_{\infty-ib}^{a-ib}\frac{x^s}{s}\,ds,\quad J_4=\int_{a+ib}^{\infty+ib}\frac{x^s}{s}\,ds.$$

Now
$$|J_3|=\left|\int_a^\infty\frac{x^{\sigma-ib}}{\sigma-ib}\,d\sigma\right|\leqslant\int_a^\infty\frac{x^\sigma}{b}\,d\sigma=-\frac{x^a}{b\log x}.$$

Similarly $|J_4|\leqslant -x^a/(b\log x)$, and the result follows in this case also.

THEOREM 16. *Let* $a>1$, $b>0$, *and* $x>1$. *Then*

$$\left|\int_{a-ib}^{a+ib}\frac{x^s}{s(s-1)}\,ds-2\pi i(x-1)\right|\leqslant\frac{2x^a}{b}.$$

Proof. It easily follows from the theorem of residues that

$$\int_{a-i\infty}^{a+i\infty}\frac{x^s}{s(s-1)}\,ds=2\pi i(x-1).$$

Also

$$\left|\int_{a+ib}^{a+i\infty}\frac{x^s}{s(s-1)}\,ds\right|=\left|\int_b^\infty\frac{x^{a+it}}{(a+it)(a-1+it)}\,dt\right|\leqslant\int_b^\infty\frac{x^a}{t^2}\,dt=\frac{x^a}{b},$$

and similarly
$$\left|\int_{a-i\infty}^{a-ib}\frac{x^s}{s(s-1)}\,ds\right|\leqslant\frac{x^a}{b}.$$

Hence the result.

THEOREM 17. *Let* $m\geqslant 3$, $a=1+1/\log(m+\frac{1}{2})$, *and*

$$S=\sum_{n=1}^\infty\left(\frac{m+\frac{1}{2}}{n}\right)^a\left|\log\frac{m+\frac{1}{2}}{n}\right|^{-1}\log n.$$

Then $S<22m(3+\log m)^2$.

Proof. Let b_n be the general term of the sum defining S. If $\frac{1}{2}m<n\leqslant 2m$, then

$$b_n\leqslant\frac{2^a\log(2m)}{|\log(m+\frac{1}{2})-\log n|}\leqslant\frac{2^a\,2m\log(2m)}{|m+\frac{1}{2}-n|},$$

and otherwise $b_n \leqslant \{(m+\frac{1}{2})/n\}^a (\log 2)^{-1} \log n$. Hence $S \leqslant S_1 + S_2$, where

$$S_1 = \sum_{n=1}^{2m} \frac{2^{a+1} m \log(2m)}{|m+\frac{1}{2}-n|} = 2^{a+2} m \log(2m) \sum_{l=1}^{m} \frac{1}{l-\frac{1}{2}}$$

$$\leqslant 16m \log(2m)(3+\log m) < 16m(3+\log m)^2,$$

$$S_2 = \frac{(m+\frac{1}{2})^a}{\log 2} \sum_{n=1}^{\infty} n^{-a} \log n = \frac{e}{\log 2}(m+\tfrac{1}{2}) \sum_{n=1}^{\infty} n^{-a} \log n,$$

and $$\sum_{n=1}^{\infty} n^{-a} \log n < \tfrac{1}{2}\log 2 + \tfrac{1}{3}\log 3 + \int_3^{\infty} x^{-a} \log x \, dx$$

$$< 1 + \int_1^{\infty} x^{-a} \log x \, dx = 1 + (a-1)^{-2}$$

$$= 1 + \log^2(m+\tfrac{1}{2}) < (3+\log m)^2,$$

so that $S_2 < 6m(3+\log m)^2$, and the result follows.

1·8. THEOREM 18. $|\psi(m)-m| \leqslant C_3 m \exp\left(-\dfrac{\sqrt{\log m}}{200}\right) \quad (m \geqslant 1)$.

Proof. Let $m \geqslant 3$,

$$a = 1 + \frac{1}{\log(m+\frac{1}{2})}, \quad \text{and} \quad b = \exp\frac{\sqrt{\log(m+\frac{1}{2})}}{100}.$$

Then, by (36), (11), and Theorems 15 and 17,

$$\left|\psi(m) + \frac{1}{2\pi i}\int_{a-ib}^{a+ib} \frac{(m+\frac{1}{2})^s}{s} \frac{\zeta'(s)}{\zeta(s)} ds\right|$$

$$= \left|\sum_{n=1}^{\infty} \Lambda(n)\left\{E\left(\frac{m+\frac{1}{2}}{n}\right) - \frac{1}{2\pi i}\int_{a-ib}^{a+ib} \frac{1}{s}\left(\frac{m+\frac{1}{2}}{n}\right)^s ds\right\}\right|$$

$$\leqslant \frac{1}{\pi b} \sum_{n=1}^{\infty} \log n \left(\frac{m+\frac{1}{2}}{n}\right)^a \left|\log \frac{m+\frac{1}{2}}{n}\right|^{-1} < \frac{22}{\pi b} m(3+\log m)^2.$$

Also, by (16), $$\frac{\eta'(s)}{\eta(s)} - \frac{\zeta'(s)}{\zeta(s)} = \frac{1}{s-1} \quad (\sigma > 1),$$

and hence, by Theorem 16,

$$\left|\frac{1}{2\pi i}\int_{a-ib}^{a+ib}\frac{(m+\frac{1}{2})^s}{s}\left(\frac{\eta'(s)}{\eta(s)}-\frac{\zeta'(s)}{\zeta(s)}\right)ds-(m-\tfrac{1}{2})\right|$$

$$\leqslant \frac{(m+\frac{1}{2})^a}{\pi b}=\frac{e}{\pi b}(m+\tfrac{1}{2}).$$

It follows that

$$\left|\psi(m)+\frac{1}{2\pi i}\int_{a-ib}^{a+ib}\frac{(m+\frac{1}{2})^s}{s}\frac{\eta'(s)}{\eta(s)}ds-(m-\tfrac{1}{2})\right|\leqslant\frac{8}{b}m(3+\log m)^2$$

$$=8m(3+\log m)^2\exp\left(-\frac{\sqrt{\log(m+\frac{1}{2})}}{100}\right)\leqslant C_4 m\exp\left(-\frac{\sqrt{\log m}}{200}\right). \tag{37}$$

Let $m>e^{250000}$ (so that $b>100$) and $a'=1-1/(10000\log b)$. Then the rectangle with corners at $a\pm ib$ and $a'\pm ib$ is contained in the set D defined in Theorem 11 and even in the subset of D to which Theorem 14 refers. Hence, by Theorem 12 and Cauchy's theorem,

$$\int_{a-ib}^{a+ib}\frac{(m+\frac{1}{2})^s}{s}\frac{\eta'(s)}{\eta(s)}ds=\int_C\frac{(m+\frac{1}{2})^s}{s}\frac{\eta'(s)}{\eta(s)}ds,$$

where C is the broken line $\{a-ib,\ a'-ib,\ a'+ib,\ a+ib\}$, and, by Theorem 14,

$$\left|\int_C\frac{(m+\frac{1}{2})^s}{s}\frac{\eta'(s)}{\eta(s)}ds\right|\leqslant 2C_2\log^3 b\left\{\int_0^b\frac{(m+\frac{1}{2})^{a'}}{|a'+it|}dt+\int_{a'}^a\frac{(m+\frac{1}{2})^\sigma}{b}d\sigma\right\}$$

$$\leqslant 2C_2\log^3 b\{(m+\tfrac{1}{2})^{a'}(2+\log b)+(m+\tfrac{1}{2})^a/b\}$$

$$=2C_2 10^{-6}\log^{\frac{3}{2}}(m+\tfrac{1}{2})$$

$$\times(m+\tfrac{1}{2})\exp\left(-\frac{\sqrt{\log(m+\frac{1}{2})}}{100}\right)\left(2+\frac{\sqrt{\log(m+\frac{1}{2})}}{100}+e\right)$$

$$\leqslant C_5 m\exp\left(-\frac{\sqrt{\log m}}{200}\right).$$

From this and (37) we obtain

$$|\psi(m)-m|\leqslant\tfrac{1}{2}+(C_4+C_5)m\exp\left(-\frac{\sqrt{\log m}}{200}\right)\quad(m>e^{250000}),$$

and the result easily follows.

1·9. Using the trivial inequalities

$$0 \leqslant \psi(m) - \vartheta(m) \leqslant \sqrt{m}\log^2 m \quad (m \geqslant 1) \tag{38}$$

(H.-W., Theorem 425), where

$$\vartheta(m) = \sum_{p \leqslant m} \log p, \tag{39}$$

we deduce from Theorem 18 that

$$|\vartheta(m) - m| \leqslant C_6 m \exp\left(-\frac{\sqrt{\log m}}{200}\right) \quad (m \geqslant 1). \tag{40}$$

It remains to prove (1).

Refinements of the prime number theorem are usually stated in terms of the 'logarithm integral'

$$\operatorname{li} x = \lim_{\epsilon \to +0}\left(\int_0^{1-\epsilon} \frac{du}{\log u} + \int_{1+\epsilon}^{x} \frac{du}{\log u}\right) \quad (x > 1).$$

For our purposes it is more convenient to use the 'logarithm sum' $\operatorname{ls} m$, which we define by

$$\operatorname{ls} m = \sum_{n=2}^{m} \frac{1}{\log n} \quad (m \geqslant 2), \ \operatorname{ls} 1 = \operatorname{ls} 0 = 0. \tag{41}$$

It is easily seen that $\operatorname{ls} m - \operatorname{li} m$ is bounded for $m \geqslant 2$.

THEOREM 19. $|\pi(m) - \operatorname{ls} m| \leqslant 3C_6 m \exp\left(-\dfrac{\sqrt{\log m}}{200}\right) \quad (m \geqslant 1).$

Proof. Putting $\Delta(n) = \vartheta(n) - n$, we have, by (40),

$$|\Delta(n)| \leqslant C_6 n \exp\left(-\frac{\sqrt{\log n}}{200}\right) \leqslant C_6 m \exp\left(-\frac{\sqrt{\log m}}{200}\right) \quad (1 \leqslant n \leqslant m).$$

Now $\pi(1) - \operatorname{ls} 1 = 0$ and, if $m \geqslant 2$, then, by (39),

$$\pi(m) = \sum_{n=2}^{m} \frac{\vartheta(n) - \vartheta(n-1)}{\log n} = \sum_{n=2}^{m} \frac{1 + \Delta(n) - \Delta(n-1)}{\log n}$$

$$= \operatorname{ls} m + \frac{\Delta(m)}{\log m} - \frac{\Delta(1)}{\log 2} + \sum_{n=2}^{m-1} \Delta(n)\left(\frac{1}{\log n} - \frac{1}{\log(n+1)}\right).$$

Hence

$$|\pi(m) - \operatorname{ls} m| \leqslant C_6 m \exp\left(-\frac{\sqrt{\log m}}{200}\right)$$

$$\times \left\{\frac{1}{\log m} + \frac{1}{\log 2} + \sum_{n=2}^{m-1}\left(\frac{1}{\log n} - \frac{1}{\log(n+1)}\right)\right\}$$

$$= \frac{2C_6}{\log 2} m \exp\left(-\frac{\sqrt{\log m}}{200}\right),$$

and the result follows, since $2/\log 2 < 3$.

CHAPTER 2

THE NUMBER OF PRIMES IN AN ARITHMETICAL PROGRESSION

2·1. We now consider

$$\pi(m;\,k,l) = \sum_{\substack{p\leqslant m \\ p\equiv l (\operatorname{mod} k)}} 1, \tag{42}$$

i.e., the number of primes in the arithmetical progression consisting of all numbers of the form $kn+l$ between 1 and m inclusive. $\pi(m;\,k,l)$ is a generalization of the function $\pi(m)$ of Chapter 1 since $\pi(m) = \pi(m;\,1,1)$. It is clear that, if $(k,l)>1$, then $\pi(m;\,k,l)$ is either 0 or 1. This case is therefore of no further interest. In the case $(k,l) = 1$ we shall obtain a result similar to (1), viz.

$$\pi(m;\,k,l) = \operatorname{ls} m/\phi(k) + O(me^{-c\sqrt{\log m}}). \tag{43}$$

Here c is again a suitable positive (absolute) constant, $\operatorname{ls} m$ is defined by (41), and $\phi(k)$ is the number of positive integers l such that $l\leqslant k$ and $(l,k) = 1$ (Euler's function, H.-W., § 5·5). It is still an unsolved problem whether (43) holds uniformly in k and l without any restriction on k. Uniformity in l alone is trivial; for $\pi(m;\,k,l) = \pi(m;\,k,l_0)$, where $l_0 = l-k[l/k]$, so that $0\leqslant l_0<k$, and corresponding to any k there are only a finite number of possible values of l_0. The aim of this chapter is to prove that, if u is any positive number, however large, then (43) holds uniformly in k and l for $k\leqslant \log^u m$.

2·2. Here is an outline of the proof. We rewrite (42) in the form

$$\pi(m;\,k,l) = \sum_{p\leqslant m} g(p), \tag{44}$$

where $g(n) = 1$ or 0 according as the condition $n\equiv l (\operatorname{mod} k)$ is or is not satisfied. We then introduce certain arithmetical functions $\chi_1, \chi_2, \ldots, \chi_{\phi(k)}$, called the characters (mod k). The precise definition of this term will be given later, but $\chi_1(n) = 1$ or 0 according as n is or is not prime to k, and we shall find that $|\chi_h(l)| = 1$ and

$$g(n) = \frac{1}{\phi(k)} \sum_{h=1}^{\phi(k)} \bar{\chi}_h(l)\,\chi_h(n), \tag{45}$$

where $\bar{\chi}_h(l)$ is the conjugate complex number to $\chi_h(l)$. Putting

$$\pi(m; \chi) = \sum_{p \leqslant m} \chi(p) \tag{46}$$

for any character χ, we deduce from (44) and (45) that

$$\pi(m; k, l) = \frac{1}{\phi(k)} \sum_{h=1}^{\phi(k)} \bar{\chi}_h(l)\,\pi(m; \chi_h). \tag{47}$$

Now $\bar{\chi}_1(l) = 1$ and

$$\pi(m) - \pi(m, \chi_1) = \sum_{p|k} 1 \leqslant \phi(k) \quad (k \leqslant m). \tag{48}$$

Hence, in view of (47) and Theorem 19, it is sufficient to prove that

$$\pi(m; \chi_h) = O(me^{-c\sqrt{\log m}}) \quad (h = 2, 3, \ldots, \phi(k)) \tag{48a}$$

uniformly for $k \leqslant \log^u m$. This will follow from properties of the functions

$$L(s, \chi) = \sum_{n=1}^{\infty} \chi(n) n^{-s} \tag{49}$$

(Dirichlet's L functions) in the same way as Theorem 19 did from those of the Riemann zeta function.

2·3. A function χ (of an integral variable) is called a *character* (mod k) if it has the following three properties:

(i) $\chi(n) = 0$ if and only if $(n, k) > 1$,

(ii) $\chi(n + k) = \chi(n)$ for every n,

and

(iii) $\chi(mn) = \chi(m)\chi(n)$ for every pair of integers m, n.

Since (iii) implies that $\chi(1) = \chi(1 \times 1) = \chi(1)\chi(1)$, and (i) that $\chi(1) \neq 0$, it follows that $\chi(1) = 1$ for every character χ. The function χ_1 defined in § 2·2 is a character (mod k), and is called the *principal character* (mod k). There are no other characters (mod 1) or (mod 2), but there is another character (mod 3), viz. the function χ for which $\chi(n) = 0, 1$, or -1 according as $n \equiv 0, 1$, or -1 (mod 3).

It is clear that $\{\chi(n)\}^q = \chi(n^q)$ for any character χ. Now, by the Fermat-Euler theorem (H.-W., Theorem 72), $n^{\phi(k)} \equiv 1$ (mod k) if $(n, k) = 1$. Hence $\{\chi(n)\}^{\phi(k)} = 1$ for any character χ (mod k) and any n prime to k. In other words, if χ is a character (mod k), and $(n, k) = 1$, then $\chi(n)$ is a $\phi(k)$th root of unity. Since a character χ (mod k) is completely determined by the values of $\chi(n)$ for $1 \leqslant n \leqslant k$, $(n, k) = 1$, it follows that there cannot be more than $\{\phi(k)\}^{\phi(k)}$ characters (mod k). We shall see later that this is a very rough estimate, and that the actual number of characters (mod k) is $\phi(k)$, but for the moment all that matters is that the number is finite.

2·4. The next theorem follows at once from the fact that, if $(k, m) = 1$, and n runs through a complete set of residues (mod k), then so does mn (H.-W., Theorem 56).

THEOREM 20. *Let $f(n + k) = f(n)$ for every n, and let $(k, m) = 1$. Then*

$$\sum_{n=1}^{k} f(mn) = \sum_{n=1}^{k} f(n).$$

THEOREM 21. *Let χ be a character* (mod k). *Then*

$$\sum_{n=1}^{k} \chi(n) = \phi(k) \text{ or } 0$$

according as χ is or is not the principal character.

Proof. The case of the principal character is trivial. Suppose, then, that χ is not the principal character. Then there is a number m such that $(k, m) = 1$ and $\chi(m) \neq 1$. Hence, by Theorem 20,

$$\sum_{n=1}^{k} \chi(n) = \sum_{n=1}^{k} \chi(mn) = \sum_{n=1}^{k} \chi(m)\,\chi(n) = \chi(m) \sum_{n=1}^{k} \chi(n),$$

and the result follows since $\chi(m) \neq 1$.

If χ_1 and χ_2 are any two characters, $\chi_1\chi_2$ denotes the function f (of an integral variable) for which $f(n) = \chi_1(n)\chi_2(n)$ for every n. It is clear that, if χ_1 and χ_2 are characters (mod k_1) and (mod k_2) respectively, then $\chi_1\chi_2$ is a character (mod k_3), where k_3 is the least common multiple of k_1 and k_2. In particular, if χ_1 and χ_2 are characters (mod k), then $\chi_1\chi_2$ is a character (mod k). Also, if χ_2 and χ_3 are different characters (mod k), and χ_1 is any character (mod k), then $\chi_1\chi_2$ and $\chi_1\chi_3$ are different characters (mod k). Since the number of characters (mod k) is finite, it follows that, if χ_1 is a character (mod k), and χ runs through all characters (mod k), then so does $\chi_1\chi$. From this we obtain the following theorem, where $\sum_{\chi (\mathrm{mod}\, k)}$ denotes a sum in which χ runs through all characters (mod k).

THEOREM 22. *Let* χ_1 *be any character* (mod k). *Then*

$$\sum_{\chi (\mathrm{mod}\, k)} \chi(n) = \sum_{\chi (\mathrm{mod}\, k)} \chi_1(n)\,\chi(n)$$

for any n.

2·5. One of our further aims is to prove the following theorem, from which it will be a short step to the determination of the number of characters (mod k) and to the proof of (45).

THEOREM 23. *Let* $m \not\equiv 1$ (mod k). *Then there is a character* χ (mod k) *such that* $\chi(m) \neq 1$.

As stated in the preface, no knowledge of group theory will be assumed in the proof. For the benefit of those readers, however, who know the very elements of this theory, I shall begin by giving an outline of the proof in terms of groups.

Suppose that G is a finite abelian group. Suppose also that, with every element A of G, there is associated a non-zero number $\chi(A)$, such that, for every pair of elements A, B of G, we have $\chi(AB) = \chi(A)\chi(B)$. Then χ is called a character of the group G.

We define the multiplication of classes of residues $(\bmod k)$ in the obvious way: If A and B are any two such classes, and the numbers a and b are members of A and B respectively, then the class of the numbers $n \equiv ab \pmod k$ (which is uniquely determined by A and B, i.e. independent of the choice of a and b) is denoted by AB. With this definition of multiplication, the set of those classes of residues $(\bmod k)$, which consist of numbers prime to k, is a finite abelian group. Let us call this group G, let χ be a character of G, and define $\chi(n)$ as follows: If $(n, k) > 1$, then $\chi(n) = 0$. If $(n, k) = 1$, and N is the class of residues $(\bmod k)$ to which n belongs, then $\chi(n) = \chi(N)$. With this definition, χ is also a character $(\bmod k)$, and it is easily seen that Theorem 23 is equivalent to the following:

THEOREM 24. *Let M be an element of G, other than the neutral element. Then there is a character χ of G such that $\chi(M) \neq 1$.*

This can be proved as follows: From those sub-groups of G, which do not contain M, we choose one with the greatest number of elements, and call it H. Then we show that the factor group G/H is cyclic. This means that, if the order of G/H is q, then there is an element B of G/H, such that $B, B^2, \ldots, B^q$ are all the elements of G/H.

For every element N of G, we now define $\chi(N)$ as follows: N belongs to exactly one element of G/H, i.e. there is exactly one integer n such that $1 \leqslant n \leqslant q$ and $N \in B^n$. We put $\chi(N) = e(n/q)$. Then it is easily seen that χ is a character of G, and that $\chi(M) \neq 1$.

2·6. In this and the next section, h^{-1} has not its usual meaning, but is defined as the integer n for which $1 \leqslant n \leqslant k$ and $hn \equiv 1$

(mod k). More generally, h^{-q} is defined as $(h^{-1})^q$. This implies, of course, that h^{-q} does not exist if $(h, k) > 1$.

I define S as the set of all integers prime to k. I define a G-set as a sub-set T of S with the following two properties:

(i) $1 \in T$.

(ii) If $m \in T$, $n \in T$ and $l \equiv mn \pmod{k}$, then $l \in T$.

It follows that any G-set consists of whole classes of residues (mod k). One G-set is said to be *greater* than another if it contains more such classes. Accordingly the least G-set is the set of the integers $n \equiv 1 \pmod{k}$, and the greatest G-set is S.

THEOREM 25. *Let h be a member of the G-set T, and let m be any integer. Then $h^m \in T$.*

Proof. If $m \geqslant 0$, the result follows immediately from the definition of the term G-set. Suppose, therefore, that $m < 0$, and let $m = -q$. Then, by the first paragraph of this section and H.-W., Theorem 72,

$$h^m = (h^{-1})^q,$$

and

$$h^{-1} \equiv h^{\phi(k)-1} \pmod{k}.$$

Hence, putting $m' = \{\phi(k) - 1\} q$, we have $m' \geqslant 0$, so that $h^{m'} \in T$, and

$$h^m \equiv h^{m'} \pmod{k}.$$

The results now follows from the fact that a G-set consists of whole classes of residues (mod k).

If $n \in S$, and T is a G-set, I define $\omega(n, T)$, the *order* of n relative to T, as the least positive integer m for which $n^m \in T$. Such positive integers m certainly exist. For instance, by H.-W., Theorem 72, $\phi(k)$ is one of them.

THEOREM 26. *Let $n \in S$, and let T be a G-set. Then $n^m \in T$ if and only if $\omega(n, T) \mid m$.*

Proof. Let

$$\omega(n, T) = q. \tag{50}$$

Then

$$n^q \in T. \tag{51}$$

Suppose, first, that $q \mid m$, and put $m/q = j$. Then, by (51) and Theorem 25, $(n^q)^j \in T$. Also

$$(n^q)^j \equiv n^{qj} \pmod{k}$$

{though not necessarily $(n^q)^j = n^{qj}$ if $j < 0$}. Since $qj = m$, it follows that $n^m \in T$.

Now suppose that $n^m \in T$. Then there are integers j, l such that

$$m = qj + l \tag{52}$$

and
$$0 \leqslant l < q. \tag{53}$$

Hence
$$n^l \equiv n^m (n^q)^{-j} \pmod{k}.$$

Also, by (51) and Theorem 25, $(n^q)^{-j} \in T$. Since T is a G-set, it follows that $n^l \in T$. Hence, by (50) and the definition of $\omega(n, T)$, l cannot be positive and less than q, so that, by (53), $l = 0$. From this and (52) it follows that $m = qj$, and from this and (50) that $\omega(n, T) \mid m$. This completes the proof.

If $h \in S$, $j \in S$, and T is a G-set, I define the statement

$$h \sim j(T)$$

(read 'h is equivalent to j relative to T') as meaning that $hj^{-1} \in T$.

THEOREM 27. *Let T be a G-set, $n \in S$, $l > 0$, $m > 0$, and*

$$\omega(n, T) = lm. \tag{54}$$

Then
$$\omega(n^l, T) = m.$$

Proof. By (54), $n^{lm} \in T$, i.e. $(n^l)^m \in T$. Hence it is sufficient to prove that, if $(n^l)^q \in T$, then $q \geqslant m$. Suppose, therefore, that $(n^l)^q \in T$. Then, since $(n^l)^q = n^{lq}$, it follows from (54) that $lq \geqslant lm$, i.e. $q \geqslant m$.

THEOREM 28. *Numbers equivalent relative to a G-set T have the same order relative to T.*

This is trivial.

If T is a G-set, and $h \in S - T$ (i.e. h is a member of S, but not of T), I define

$$\text{ext}\,(T, h)$$

as the set of those integers which are equivalent to powers of h relative to T. A set U is called a *simple extension* of a G-set T if and only if there is a number h in $S - T$ such that $U = \text{ext}\,(T, h)$.

LEMMA 1. *Any simple extension of a G-set T is a G-set, contains T, and is greater than T.*

This, too, is trivial.

LEMMA 2. *Let T and U be G-sets, $n \in S$, and $T \subset U$. Then*

$$\omega(n, U) \mid \omega(n, T).$$

This follows from Theorem 26.

2·7. LEMMA 3. *Let T be a G-set other than S, let h be a member of S of greatest order relative to T, and let*

$$U = \text{ext}\,(T, h). \tag{55}$$

Then either $U = S$, or there is a number j in $S - U$ such that

$$\omega(j, U) = \omega(j, T).$$

Proof. Suppose $U \neq S$. Let l be any member of $S - U$, and put

$$\omega(l, U) = q. \tag{56}$$

Then $$l^q \in U.$$

Hence, by (55), there is an integer m such that

$$l^q \sim h^m(T). \tag{57}$$

There are integers n, r such that

$$m = nq + r \tag{58}$$

and $$0 \leqslant r < q. \tag{59}$$

Let $$j = lh^{-n}. \tag{60}$$

Then, by (55), $$j \sim l(U),$$

and hence, by Theorem 28 and (56),

$$\omega(j, U) = q. \tag{61}$$

Also, by (60), (57) and (58),

$$j^q \sim h^r(T). \tag{62}$$

Now, by Lemma 2, (55) and (61),

$$q \mid \omega(j, T),$$

so that we may put $$\omega(j, T) = qq_1. \tag{63}$$

From this and Theorem 27 it follows that $\omega(j^q, T) = q_1$. Hence, by (62) and Theorem 28,

$$\omega(h^r, T) = q_1,$$

which implies that $$h^{rq_1} \in T.$$

From this it follows that, if $r > 0$, then $rq_1 \geqslant \omega(h, T)$, and hence, by (63) and (59), $\omega(j, T) > \omega(h, T)$, which contradicts the hypothesis that h is a member of S of greatest order relative to T. Thus, by (59), $r = 0$, and hence, by (62), $j^q \in T$, which implies that $q \geqslant \omega(j, T)$, and so, by (61),

$$\omega(j, U) \geqslant \omega(j, T).$$

On the other hand, since $T \subset U$, it is trivial that $\omega(j, U) \leqslant \omega(j, T)$, and the result follows.

LEMMA 4. *Let T be a G-set, $m \in S - T$, and suppose that S is not a simple extension of T. Then T has a simple extension V such that $m \in S - V$.*

Proof. Let h be a member of S of greatest order relative to T, and let U be defined by (55). If m is not in U, the result is trivial, for we need only take $V = U$. Suppose, therefore, that $m \in U$. Then, by Lemma 3, there is a number j in $S - U$ such that

$$\omega(j, U) = \omega(j, T). \tag{64}$$

Let
$$V = \text{ext}(T, j),$$

and suppose, if possible, that $m \in V$. Then there is an integer n such that

$$m \sim j^n(T). \tag{64a}$$

Since $T \subset U$, it follows that $m \sim j^n(U)$, and since $m \in U$, it now follows that $j^n \in U$. Hence, by (64) and Theorem 26, $j^n \in T$. From this and (64a) it follows that $m \in T$, which contradicts the hypothesis $m \in S - T$. Thus m is not in V, i.e. $m \in S - V$.

LEMMA 5. *Let T be a G-set, $h \in S$, and*

$$h^{j_1} \sim h^{j_2}(T).$$

Then
$$\omega(h, T) \mid j_1 - j_2.$$

This follows easily from Theorem 26.

Proof of Theorem 23. The case $(m, k) > 1$ is trivial. So let $(m, k) = 1$, i.e. $m \in S$. Since $m \not\equiv 1 \pmod{k}$, there is at least one G-set which does not contain m, namely the set of the integers $n \equiv 1 \pmod{k}$. Hence there is a greatest such G-set, which we may call T. Then, by Lemmas 4 and 1, S is a simple extension of T. This means that there is a number h in $S - T$ such that

$$S = \text{ext}(T, h).$$

Hence, for every n in S, there is an integer j such that

$$n \sim h^j(T),$$

and it follows from Lemma 5 that all such integers j are in the same class of residues $(\text{mod}\, q)$, where

$$q = \omega(h, T).$$

Hence $e(j/q)$ has the same value for all such integers j. Denote this value by $\chi(n)$. Then $\chi(n)$ is defined for every n in S. If n is not in S, let $\chi(n) = 0$. Then it is easily seen that χ is a character (mod k). Also, if

$$m \sim h^j(T),$$

then, since m is not in T, neither is h^j. Hence, by Theorem 26,

$$q \nmid j, \text{ i.e. } e(j/q) \neq 1, \text{ i.e. } \chi(m) \neq 1.$$

2·8. THEOREM 29. *Let* $n \not\equiv 1 \pmod{k}$. *Then* $\sum_{\chi \pmod{k}} \chi(n) = 0$.

Proof. By Theorem 23, there is a character $\chi_1 \pmod{k}$ such that $\chi_1(n) \neq 1$. Now, by Theorem 22,

$$\sum_{\chi \pmod{k}} \chi(n) = \chi_1(n) \sum_{\chi \pmod{k}} \chi(n),$$

and the result follows.

THEOREM 30. *There are exactly* $\phi(k)$ *characters* (mod k).

Proof. It was shown in § 2·3 that $\chi(1) = 1$ for every character χ. Hence, by Theorems 29 and 21,

$$\sum_{\chi \pmod{k}} 1 = \sum_{\chi \pmod{k}} \chi(1) = \sum_{n=1}^{k} \sum_{\chi \pmod{k}} \chi(n) = \sum_{\chi \pmod{k}} \sum_{n=1}^{k} \chi(n) = \phi(k).$$

THEOREM 31. $\sum_{\chi \pmod{k}} \chi(n) = \phi(k)$ *or* 0 *according as the condition* $n \equiv 1 \pmod{k}$ *is or is not satisfied.*

This follows from Theorems 30 and 29 on noting that, by § 2·3, $\chi(n) = 1$ for any character $\chi \pmod{k}$ if $n \equiv 1 \pmod{k}$.

The next theorem is equivalent to (45).

THEOREM 32. *Let* $(k, l) = 1$. *Then* $\sum_{\chi \pmod{k}} \bar{\chi}(l)\chi(n) = \phi(k)$ *or* 0 *according as the condition* $n \equiv l \pmod{k}$ *is or is not satisfied.*

Proof. Since $(k, l) = 1$, there is an integer m such that

$$lm \equiv n \pmod{k}.$$

Hence, by § 2·3,

$$\bar{\chi}(l)\chi(n) = \bar{\chi}(l)\chi(l)\chi(m) = |\chi(l)|^2\chi(m) = \chi(m),$$

and hence, by Theorem 31,

$$\sum_{\chi (\bmod k)} \bar{\chi}(l)\chi(n) = \sum_{\chi (\bmod k)} \chi(m) = \begin{cases} \phi(k) & (m \equiv 1 \pmod{k}), \\ 0 & \text{(otherwise)}, \end{cases}$$

and it is clear that $m \equiv 1 \pmod{k}$ if and only if $n \equiv l \pmod{k}$.

2·9. We now turn to the investigation of Dirichlet's L functions. The following simple theorem will be useful in this connexion.

THEOREM 33. *Let χ be a character* (mod k), *but not the principal character. Then, for any positive integer n,*

$$\left| \sum_{m=1}^{n} \chi(m) \right| \leqslant \tfrac{1}{2}k.$$

Proof. Let $a_0 = 0$ and

$$a_n = \sum_{m=1}^{n} \chi(m) \quad (n = 1, 2, \ldots).$$

Then, since $|\chi(m)| \leqslant 1$ for any m,

$$|a_n - a_l| \leqslant |n - l| \quad (n \geqslant 0, l \geqslant 0).$$

Also, by Theorem 21 and § 2·3(ii), $a_l = 0$ if l is a positive multiple of k. Thus, taking $l = k[(n/k) + \frac{1}{2}]$, we obtain

$$|a_n| = |a_n - a_l| \leqslant |n - l| \leqslant \tfrac{1}{2}k,$$

which proves the theorem.

We now develop a theory of Dirichlet's L functions, defined by (49), similar to that of the Riemann zeta function given in Chapter 1. In general, we shall confine ourselves here to L functions formed with non-principal characters. For this there are two reasons. Firstly, we do not need the principal characters any more, as they do not occur in (48*a*). Secondly, the L functions formed with principal characters are not essentially different from the zeta function. More precisely, if χ is the principal character (mod k), then

$$L(s,\chi) = \zeta(s)\prod_{p|k}(1-p^{-s}).$$

This is easily proved for $\sigma > 1$, and may be taken as the definition of $L(s,\chi)$ for $\sigma \leqslant 1$.

THEOREM 34. *Let χ be a non-principal character* (mod k), *and let $\sigma > 0$ and $1 \leqslant n_1 \leqslant n_2$. Then*

$$\left|\sum_{n=n_1}^{n_2} \chi(n)\, n^{-s}\right| \leqslant k\frac{|s|}{\sigma} n_1^{-\sigma}.$$

Proof. With a_n defined as in the proof of Theorem 33, we have

$$\sum_{n=n_1}^{n_2} \chi(n)\, n^{-s} = \sum_{n=n_1}^{n_2} (a_n - a_{n-1})\, n^{-s}$$

$$= \sum_{n=n_1}^{n_2} a_n\{n^{-s} - (n+1)^{-s}\} - a_{n_1-1} n_1^{-s} + a_{n_2}(n_2+1)^{-s}.$$

Now

$$|n^{-s} - (n+1)^{-s}| = \left|\int_n^{n+1} s x^{-s-1}\, dx\right| \leqslant |s| \int_n^{n+1} x^{-\sigma-1}\, dx$$

and, by Theorem 33, $|a_n| \leqslant \frac{1}{2}k$. Hence

$$\left|\sum_{n=n_1}^{n_2} \chi(n)\, n^{-s}\right| \leqslant \tfrac{1}{2}k\left\{|s| \int_{n_1}^{n_2+1} x^{-\sigma-1}\, dx + n_1^{-\sigma} + (n_2+1)^{-\sigma}\right\}$$

$$= \tfrac{1}{2}k\left\{\frac{|s|}{\sigma}\{n_1^{-\sigma} - (n_2+1)^{-\sigma}\} + n_1^{-\sigma} + (n_2+1)^{-\sigma}\right\} \leqslant k\frac{|s|}{\sigma} n_1^{-\sigma}.$$

THEOREM 35. *Let χ be a non-principal character* (mod k). *Then $L(s,\chi)$ is regular for $\sigma > 0$, and*

$$|L(s,\chi)| \leqslant k|s|/\sigma \quad (\sigma > 0).$$

Proof. It follows from Theorem 34 that $\sum_{n=1}^{\infty} \chi(n)\, n^{-s}$ converges locally uniformly in the half-plane $\sigma > 0$. Hence $L(s,\chi)$ is regular there. The last part of the theorem follows from Theorem 34 on taking $n_1 = 1$ and letting $n_2 \to \infty$.

We note that, by Theorem 35,

$$|L(1,\chi)| \leqslant k \tag{65}$$

if χ is a non-principal character $(\bmod k)$. The following result, which is more precise than (65) when k is large, will be used later.

THEOREM 36. *Let χ be a non-principal character* $(\bmod k)$. *Then*

$$|L(1,\chi)| \leqslant 2+\log k.$$

Proof. Using Theorem 34 with $s=1$, $n_1 = k$, and letting $n_2 \to \infty$, we obtain

$$\left|\sum_{n=k}^{\infty} \chi(n)\, n^{-1}\right| \leqslant 1.$$

Hence, by (49),

$$|L(1,\chi)| \leqslant \left|\sum_{n=1}^{k-1} \chi(n)\, n^{-1}\right| + 1 \leqslant 1 + \sum_{n=1}^{k-1} n^{-1} \leqslant 2+\log k.$$

For any character χ, by (49) and Theorem 2,

$$L(s,\chi) = \prod_p \{1-\chi(p)p^{-s}\}^{-1} \quad (\sigma > 1) \tag{66}$$

which implies that $\quad L(s,\chi) \neq 0 \quad (\sigma > 1). \qquad (67)$

Also

$$\frac{L'(s,\chi)}{L(s,\chi)} = -\sum_{n=1}^{\infty} \Lambda(n)\,\chi(n)\, n^{-s} \quad (\sigma > 1). \tag{68}$$

This follows from (66) in the same way as (11) from Theorem 3.

THEOREM 37. *Let $u > 1$, and let χ be any character. Then*

$$\mathbf{R}\left\{3\frac{\zeta'(u)}{\zeta(u)} + 4\frac{L'(u+iv,\chi)}{L(u+iv,\chi)} + \frac{L'(u+2iv,\chi^2)}{L(u+2iv,\chi^2)}\right\} \leqslant 0$$

and

$$|\zeta^3(u)\, L^4(u+iv,\chi)\, L(u+2iv,\chi^2)| \geqslant 1.$$

The proof is similar to that of Theorem 6, but we now use (68) as well as (11), (66) as well as Theorem 3, take

$$a_n = 3+4\chi(n)n^{-iv}+\chi^2(n)n^{-2iv},$$

and use Theorem 5 only in the case $(n, k) = 1$ (assuming that χ is a character (mod k)), the case $(n, k) > 1$ being trivial.

A character χ is said to be *real* if $\chi(n)$ is real for every n. We note that, if χ is a real character, then χ^2 is a principal character and vice versa.

THEOREM 38. *Let $u > 1$, and let χ be any real character. Then*

$$\mathbf{R}\left\{3\frac{\zeta'(u)}{\zeta(u)}+4\frac{L'(u+iv,\chi)}{L(u+iv,\chi)}+\frac{\zeta'(u+2iv)}{\zeta(u+2iv)}\right\}\leqslant 0$$

and

$$|\zeta^3(u)L^4(u+iv,\chi)\zeta(u+2iv)|\geqslant 1.$$

The proof is similar to that of Theorem 37, but we now take $a_n = 3+4\chi(n)n^{-iv}+n^{-2iv}$.

2·10. We found in Theorem 11 that the zeta function has no zeros 'too near' the line $\sigma = 1$. A similar result is true of Dirichlet's L functions, and can be proved in a similar way, with the following important difference. It is trivial that the zeta function has no zeros too near the point $s = 1$, as it has a pole *at* this point. The L functions formed with non-principal characters, however, have no pole at this point, and though they have no zero there either, even this is not trivial. The method of the proof of Theorem 11 is applicable, almost without modification, to L functions formed with non-real characters. A similar method works with L functions formed with real non-principal characters as long as we restrict ourselves to non-real zeros, but an entirely different method must be used to show that such functions have no zeros *on the real axis* at or near the point $s = 1$.

Instead of the number t^* defined by (15), we now use

$$\tau = \max(|t|, k, 100). \tag{69}$$

THEOREM 39. *Let χ be a non-real character* (mod k). *Then*

$$L(s,\chi)\neq 0 \quad (\sigma > 1-1/(4000\log\tau)).$$

Proof. Suppose that this is not so. Then there are numbers σ_0 and t_0 such that

$$L(\sigma_0+it_0,\chi)=0 \tag{70}$$

and

$$\sigma_0 > 1-1/(4000\log\tau_0), \tag{71}$$

where

$$\tau_0 = \max(|t_0|, k, 100). \tag{72}$$

It follows from (67) and (70) that

$$\sigma_0 \leqslant 1. \tag{73}$$

Now, if $1<u<2$ and $|s-u-it_0|\leqslant \frac{1}{4}$, then $|s|\leqslant 3+\tau_0$ and $\sigma > \frac{3}{4}$, so that, by Theorem 35 and (72),

$$|L(s,\chi)| \leqslant k(4+\tfrac{4}{3}\tau_0) \leqslant \tfrac{3}{2}\tau_0^2 \quad (1<u<2, |s-u-it_0|\leqslant \tfrac{1}{4}). \tag{74}$$

Similarly, since χ^2 is a non-principal character (mod k),

$$|L(s,\chi^2)| \leqslant 3\tau_0^2 \quad (1<u<2, |s-u-2it_0|\leqslant \tfrac{1}{4}). \tag{75}$$

Now let

$$u = 1+1/(800\log\tau_0) \tag{76}$$

and

$$f(z) = \eta^3(u+\tfrac{1}{8}z)L^4(u+it_0+\tfrac{1}{8}z,\chi)L(u+2it_0+\tfrac{1}{8}z,\chi^2), \tag{77}$$

where $\eta(s)$ is defined by (16). Then, by (23), (74), (75) and (72),

$$|f(z)| \leqslant (\tfrac{4}{3})^3(\tfrac{3}{2}\tau_0^2)^4 3\tau_0^2 < \tau_0^{11} \quad (|z|\leqslant 2). \tag{78}$$

Also, putting

$$g(w) = \zeta^3(u+w)L^4(u+it_0+w,\chi)L(u+2it_0+w,\chi^2), \tag{79}$$

we have, by (77) and (16), $f(z) = (u-1+\frac{1}{8}z)^3 g(\frac{1}{8}z)$, which implies that

$$f(0) = (u-1)^3 g(0) \tag{80}$$

and

$$\frac{f'(0)}{f(0)} = \frac{3}{8(u-1)} + \frac{1}{8}\frac{g'(0)}{g(0)}. \tag{81}$$

These two formulae look like (28) and (29), but have a different meaning. Now, by (79) and Theorem 37, $|g(0)| \geqslant 1$ and $\mathbf{R}\{g'(0)/g(0)\} \leqslant 0$. Hence, by (76), (80), and (81),

$$|f(0)| \geqslant (800 \log \tau_0)^{-3} \tag{82}$$

and
$$\mathbf{R}\{f'(0)/f(0)\} \leqslant 300 \log \tau_0. \tag{83}$$

Also (cf. the proof of Theorem 11) $\tau_0/\log \tau_0 > 20$, and hence, by (82), $|f(0)| > (40\tau_0)^{-3} > \tau_0^{-11/2}$. From this and (78) it follows that

$$|f(z)/f(0)| \leqslant \tau_0^{33/2} \quad (|z| \leqslant 2). \tag{84}$$

Now, by (70) and (77), $f(z)$ has a zero at $z = 8(\sigma_0 - u)$, and, denoting the order of this zero by h, we have (33). Putting $M = \frac{33}{2} \log \tau_0$ and $a = 8(u - \sigma_0)$, we can now verify that all the hypotheses of Theorem 9 are satisfied. In fact, $f(z)$ is regular for $|z| \leqslant 2$ by (77), (76), (16), Theorem 1, and Theorem 35; $|f(z)/f(0)| \leqslant e^M$ $(|z| \leqslant 2)$ by (84); $0 < a \leqslant 1$ by (76), (73), (71), and (72); (14) follows from (77), (16), (10), and (67) since $u > 1$, and we have just seen that $f(z)$ has a zero of order h at $z = -a$. Hence, by Theorem 9, (83), and (33),

$$0 \leqslant 2M - \frac{4}{a} + 300 \log \tau_0 = 333 \log \tau_0 - \frac{1}{2(u - \sigma_0)}.$$

From this and (76) and (71) it follows that

$$333 \geqslant \frac{1}{2(u - \sigma_0) \log \tau_0} > \frac{1}{2(800^{-1} + 4000^{-1})} = \frac{1000}{3}.$$

This is a contradiction, and so Theorem 39 is proved.

The narrow margin by which this contradiction was obtained reminds me of the story of the Scotsman who looked suspiciously at his change, and when asked if it was not enough, said: 'Yes, but only just.'

THEOREM 40. *Let χ be a real non-principal character* (mod k). *Then*

$$L(s,\chi) \neq 0 \quad (\sigma > 1 - 1/(8000 \log \tau), t \neq 0).$$

Proof. Suppose that this is not so. Then there are numbers σ_0 and t_0 such that (70) holds, that

$$t_0 \neq 0, \tag{85}$$

and that
$$\sigma_0 > 1 - 1/(8000 \log \tau_0), \tag{86}$$

with τ_0 again defined by (72). (73) and (74) hold as before, but χ^2 is now a principal character, and so we avoid $L(s,\chi^2)$.

Suppose, first, that

$$|t_0| > 1/(400 \log k_0), \tag{87}$$

where
$$k_0 = \max(k, 100). \tag{88}$$

Then we put
$$u = 1 + 1/(1600 \log \tau_0) \tag{89}$$

and
$$f(z) = \eta^3(u + \tfrac{1}{8}z) L^4(u + it_0 + \tfrac{1}{8}z, \chi) \eta(u + 2it_0 + \tfrac{1}{8}z) \tag{90}$$

instead of (76) and (77) respectively. It follows from (16) and (9) that

$$|\eta(s)| \leqslant 1 + \tfrac{4}{3}|s||s-1| \leqslant 1 + \tfrac{4}{3}(|t|+2)^2 \quad (\tfrac{3}{4} \leqslant \sigma \leqslant 2).$$

Hence, by (89) and (72),

$$|\eta(u + 2it_0 + \tfrac{1}{8}z)| \leqslant 1 + \tfrac{4}{3}(2|t_0|+3)^2 \leqslant 6\tau_0^2 \quad (|z| \leqslant 2). \tag{91}$$

By (90), (23), (74), (91), and (72),

$$|f(z)| \leqslant (\tfrac{4}{3})^3 (\tfrac{3}{2}\tau_0^2)^4 6\tau_0^2 < \tau_0^{11} \quad (|z| \leqslant 2). \tag{92}$$

Now let

$$g(w) = \zeta^3(u+w) L^4(u + it_0 + w, \chi) \zeta(u + 2it_0 + w) \tag{93}$$

(instead of (79)). Then, by (90) and (16),

$$f(z) = (u - 1 + \tfrac{1}{8}z)^3 (u - 1 + 2it_0 + \tfrac{1}{8}z) g(\tfrac{1}{8}z),$$

which implies that

$$f(0) = (u-1)^3 (u - 1 + 2it_0) g(0) \tag{94}$$

and
$$\frac{f'(0)}{f(0)} = \frac{3}{8(u-1)} + \frac{1}{8(u-1+2it_0)} + \frac{1}{8}\frac{g'(0)}{g(0)}. \tag{95}$$

Now, by (93) and Theorem 38, $|g(0)| \geqslant 1$ and $\mathbf{R}\{g'(0)/g(0)\} \leqslant 0$, and it follows from (87), (88), and (72) that $|t_0| > 1/(400 \log \tau_0)$. Hence, by (94), (89), and (95),

$$|f(0)| \geqslant (u-1)^4 = (1600 \log \tau_0)^{-4} \tag{96}$$

and

$$\mathbf{R}\frac{f'(0)}{f(0)} \leqslant \frac{3}{8(u-1)} + \frac{1}{8}\frac{u-1}{(u-1)^2 + (200 \log \tau_0)^{-2}} = 603\tfrac{1}{17} \log \tau_0. \tag{97}$$

Also, as shown in the proof of Theorem 39, $\tau_0/\log \tau_0 > 20$. Hence, by (96), $|f(0)| > (80\tau_0)^{-4} > \tau_0^{-8}$. From this and (92) it follows that

$$|f(z)/f(0)| < \tau_0^{19} \quad (|z| \leqslant 2). \tag{98}$$

We now define h and a as in the proof of Theorem 39, put $M = 19 \log \tau_0$ instead of $\frac{33}{2} \log \tau_0$, and verify the hypotheses of Theorem 9 by the same argument as in the proof of Theorem 39, but use (90), (89), (98), and (86) instead of (77), (76), (84), and (71) respectively. We then obtain from Theorem 9, (97), and (33) that

$$0 \leqslant 2M \frac{4}{a} + 604 \log \tau_0 = 642 \log \tau_0 \frac{1}{2(u-\sigma_0)}.$$

From this and (89) and (86) it follows that

$$642 \geqslant \frac{1}{2(u-\sigma_0)\log \tau_0} > \frac{1}{2(1600^{-1} + 8000^{-1})} = \frac{2000}{3}.$$

This is a contradiction.

Now suppose that

$$|t_0| \leqslant 1/(400 \log k_0). \tag{99}$$

Then we put

$$u = 1 + 1/(100 \log k_0), \tag{100}$$

$$f(z) = \eta(u + \tfrac{1}{8}z) L(u + \tfrac{1}{8}z, \chi), \tag{101}$$

and

$$g(w) = \zeta(u+w) L(u+w, \chi) \tag{102}$$

instead of (89), (90), and (93) respectively. By (101), (102), and (16), $f(z) = (u - 1 + \frac{1}{8}z) g(\frac{1}{8}z)$, which implies that

$$f(0) = (u-1) g(0) \tag{103}$$

and

$$\frac{f'(0)}{f(0)} = \frac{1}{8(u-1)} + \frac{1}{8}\frac{g'(0)}{g(0)}. \tag{104}$$

Now obviously $1 < u < 21/20$, so that, by (23),

$$|\eta(u+\tfrac{1}{8}z)| < \tfrac{4}{3} \quad (|z| \leqslant 2).$$

Also, by the last formula preceding (23), $|s|/\sigma < 31/30$ if $|s-u| \leqslant \frac{1}{4}$, so that, by Theorem 35,

$$|L(u+\tfrac{1}{8}z, \chi)| < \tfrac{31}{30}k \quad (|z| \leqslant 2).$$

It therefore follows from (101) that

$$|f(z)| < \tfrac{4}{3} \times \tfrac{31}{30}k < \tfrac{7}{5}k \quad (|z| \leqslant 2). \tag{105}$$

Now, by (102), Theorem 3, and (66),

$$g(0) = \zeta(u) L(u, \chi) = \prod_p \{(1-p^{-u})(1-\chi(p)p^{-u})\}^{-1},$$

and $\chi(p)$, being real, can only assume the values 0, 1, and -1, so that $0 < (1-p^{-u})(1-\chi(p)p^{-u}) < 1$. Hence $g(0) > 1$, and hence, by (103) and (100),

$$f(0) > u - 1 = (100 \log k_0)^{-1}. \tag{106}$$

Also, by (102), (11) and (68),

$$\frac{g'(0)}{g(0)} = \frac{\zeta'(u)}{\zeta(u)} + \frac{L'(u,\chi)}{L(u,\chi)} = -\sum_{n=1}^{\infty} \Lambda(n)\{1+\chi(n)\} n^{-u} \leqslant 0,$$

and hence, by (104) and (100),

$$\frac{f'(0)}{f(0)} \leqslant \frac{1}{8(u-1)} = \frac{25}{2} \log k_0. \tag{107}$$

Now, by (88), $k \leqslant k_0$ and $k_0/\log k_0 \geqslant 100/\log 100 > 20$. Hence, by (105) and (106),

$$|f(z)/f(0)| < 7k \times 20 \log k_0 < 7k_0^2 < k_0^{5/2} \quad (|z| \leqslant 2). \tag{108}$$

Putting

$$M = \tfrac{5}{2} \log k_0, \; a = 8(-u+\sigma_0+it_0), \text{ and } b = 8(-u+\sigma_0-it_0),$$

we now find that all the hypotheses of Theorem 10 are satisfied. In fact, $f(z)$ is regular for $|z| \leqslant 2$ by (101), (16), Theorem 1, and Theorem 35; $|f(z)/f(0)| \leqslant e^M$ $(|z| \leqslant 2)$ by (108); $|a| \leqslant 1$ and $|b| \leqslant 1$ by (100), (86), (73), and (99); $a \neq b$ by (85); $f(a) = 0$ by (101) and (70), and it follows that $f(b) = 0$ since a and b are conjugate complex numbers and χ is real; finally (14) follows from (101), (16), (10), and (67) since $u > 1$. Hence, by Theorem 10 and (107),

$$0 \leqslant \frac{25}{2}\log k_0 + 2M + \mathbf{R}\frac{1}{a} + \mathbf{R}\frac{1}{b} = \frac{35}{2}\log k_0 + \frac{1}{4}\frac{-u+\sigma_0}{(u-\sigma_0)^2 + t_0^2} \quad (109)$$

Now, by (99), (72), and (88), $\tau_0 = k_0$, and hence, by (73), (100), and (86),

$$100^{-1} \leqslant (u-\sigma_0)\log k_0 < 100^{-1} + 8000^{-1}.$$

From this and (109) and (99) it follows that

$$70 \geqslant \frac{(u-\sigma_0)\log k_0}{\{(u-\sigma_0)\log k_0\}^2 + (t_0 \log k_0)^2} \geqslant \frac{100^{-1}}{(100^{-1} + 8000^{-1})^2 + 400^{-2}}$$

$$= \frac{640000}{81^2 + 400} = \frac{640000}{6961} > \frac{640}{7}.$$

This is again a contradiction, and so the theorem is proved.

2·11. It remains to be shown that L functions formed with real non-principal characters have no zeros on the real axis at or 'too near' the point $s = 1$. The following elementary theorem is useful in this connexion.

THEOREM 41. *Let $f(n)$ be multiplicative, and let $\prod_p \sum_{m=0}^{\infty} |f(p^m)|$ converge. Then*

$$\sum_{n=1}^{\infty} f(n) = \prod_p \sum_{m=0}^{\infty} f(p^m).$$

Proof. In virtue of Theorem 2, it is sufficient to prove that $\sum_{n=1}^{\infty} |f(n)|$ converges. We assume that $f(n)$ is not identically zero.

Hence $$\sum_{m=0}^{\infty} |f(p^m)| \geqslant |f(1)| = f(1) = 1$$

for any p, and hence, putting

$$\prod_p \sum_{m=0}^{\infty} |f(p^m)| = A,$$

we have $$\prod_{p \leqslant k} \sum_{m=0}^{\infty} |f(p^m)| \leqslant A$$

for any k. Now obviously

$$\prod_{p \leqslant k} \sum_{m=0}^{[\log k/\log p]} |f(p^m)| \leqslant \prod_{p \leqslant k} \sum_{m=0}^{\infty} |f(p^m)|$$

and $$\prod_{p \leqslant k} \sum_{m=0}^{[\log k/\log p]} |f(p^m)| = \sum_{n \in \alpha} |f(n)|,$$

where α is the aggregate of all positive integers not divisible by prime powers greater than k. This aggregate contains the integers from 1 to k, so that

$$\sum_{n=1}^{k} |f(n)| \leqslant \sum_{n \in \alpha} |f(n)|.$$

It follows that $$\sum_{n=1}^{k} |f(n)| \leqslant A.$$

Since this holds for every k, we deduce that $\sum_{n=1}^{\infty} |f(n)|$ converges, and the theorem is proved.

THEOREM 42. *Corresponding to any real character χ, there is a function $f(n)$, such that*

$$f(n) \geqslant 0 \quad (n = 1, 2, \ldots), \tag{110}$$

$$f(n^2) \geqslant 1 \quad (n = 1, 2, \ldots), \tag{111}$$

and $$\zeta(s) L(s, \chi) = \sum_{n=1}^{\infty} f(n) n^{-s} \quad (\sigma > 1). \tag{112}$$

Proof. By Theorem 3 and (66),

$$\zeta(s) L(s, \chi) = \prod_p \phi(s, p) \quad (\sigma > 1),$$

where $$\phi(s, p) = ((1 - p^{-s})\{1 - \chi(p) p^{-s}\})^{-1}.$$

Now the only values which $\chi(p)$ can have are 0, 1, and -1. If $\chi(p) = 0$,

$$\phi(s,p) = (1-p^{-s})^{-1} = \sum_{m=0}^{\infty} p^{-ms} \quad (\sigma > 0).$$

If $\chi(p) = 1$,

$$\phi(s,p) = (1-p^{-s})^{-2} = \sum_{m=0}^{\infty} (m+1)p^{-ms} \quad (\sigma > 0).$$

If $\chi(p) = -1$,

$$\phi(s,p) = (1-p^{-2s})^{-1} = \sum_{m=0}^{\infty} p^{-2ms} \quad (\sigma > 0).$$

Thus, putting $f(1) = 1$, $f(p^q) = 1$ if $\chi(p) = 0$, $f(p^q) = q+1$ if $\chi(p) = 1$, $f(p^q) = 1$ if $\chi(p) = -1$ and q is even, and $f(p^q) = 0$ if $\chi(p) = -1$ and q is odd, we have

$$\phi(s,p) = \sum_{m=0}^{\infty} f(p^m)p^{-ms} \quad (\sigma > 0)$$

for any p, and hence

$$\zeta(s)L(s,\chi) = \prod_p \sum_{m=0}^{\infty} f(p^m)p^{-ms} \quad (\sigma > 1). \tag{113}$$

So far $f(n)$ has only been defined when n is 1 or a prime power. When

$$n = p_1^{q_1} p_2^{q_2} \dots p_l^{q_l},$$

where $p_1, p_2, \dots, p_l$ are distinct primes, we put

$$f(n) = \prod_{h=1}^{l} f(p_h^{q_h}).$$

Then $f(n)$ is defined for every positive n, is clearly multiplicative, and satisfies (110) and (111). Also $\prod_p \sum_{m=0}^{\infty} |f(p^m)p^{-ms}|$ converges for $\sigma > 1$, viz. to $\zeta(\sigma)L(\sigma,\chi)$, by (113) with σ instead of s. Hence (112) follows from (113) and Theorem 41, the latter with $f(n)n^{-s}$ instead of $f(n)$. This completes the proof.

4

THEOREM 43. *Corresponding to any two real characters* χ_0 *and* χ, *there is a function* $f(n)$, *such that* (110) *holds, that*

$$f(1) = 1, \tag{114}$$

and that

$$\zeta(s)\,L(s,\chi_0)\,L(s,\chi)\,L(s,\chi_0\chi) = \sum_{n=1}^{\infty} f(n)\,n^{-s} \quad (\sigma > 1). \tag{115}$$

The proof is similar to that of Theorem 42, and may therefore be left to the reader.

THEOREM 44. *Let* χ *be any real non-principal character. Then* $L(1,\chi) \neq 0$.

Proof. Putting $g(s) = \zeta(s)\,L(s,\chi)$, we have, by Theorem 42,

$$g(s) = \sum_{n=1}^{\infty} f(n)\,n^{-s} \quad (\sigma > 1),$$

where $f(n)$ satisfies (110) and (111), which means that $f(n) \geqslant a_n$ $(n = 1, 2, \ldots)$, where $a_n = 1$ or 0 according as n is or is not a square. Hence

$$(-1)^m g^{(m)}(2) = \sum_{n=1}^{\infty} f(n)\,n^{-2}\log^m n \geqslant \sum_{n=1}^{\infty} a_n\,n^{-2}\log^m n$$

$$= \sum_{l=1}^{\infty} (l^2)^{-2}\log^m (l^2) = (-2)^m \zeta^{(m)}(4)$$

$$(m = 0, 1, 2, \ldots).$$

Now suppose, if possible, that $L(1,\chi) = 0$. Then, by Theorems 1 and 35, $g(s)$ is regular for $\sigma > 0$. Hence, for any s for which $\frac{1}{2} < s \leqslant 2$,

$$g(s) = \sum_{m=0}^{\infty} (-1)^m \frac{g^{(m)}(2)}{m!}(2-s)^m \geqslant \sum_{m=0}^{\infty} (-2)^m \frac{\zeta^{(m)}(4)}{m!}(2-s)^m = \zeta(2s).$$

This is impossible, since $g(s) \to g(\frac{1}{2})$ and $\zeta(2s) \to \infty$ when $s \to \frac{1}{2} + 0$.

2·12. Before proceeding further, we have to introduce the term 'equivalent characters'. Two characters χ_0 and χ are said to be *equivalent* if $\chi_0(n) = \chi(n)$ for every n for which neither $\chi_0(n)$ nor $\chi(n)$ is 0. It follows at once that, if χ_0 and χ are any two non-equivalent *real* characters, then $\chi_0\chi$ is not a principal character.

THEOREM 45. *Any two L functions formed with equivalent non-principal characters have the same zeros in the half-plane* $\sigma > 0$.

Proof. Let χ_0 and χ be any two equivalent non-principal characters, (mod k_0) and (mod k) respectively, and let

$$f_0(s) = \prod_{p \mid k_0 k} \{1 - \chi_0(p) p^{-s}\}, \quad f(s) = \prod_{p \mid k_0 k} \{1 - \chi(p) p^{-s}\}.$$

Then $\chi_0(p) = \chi(p)$ $(p \nmid k_0 k)$, and hence, by (66),

$$\begin{aligned} f_0(s) L(s, \chi_0) &= \prod_{p \nmid k_0 k} \{1 - \chi_0(p) p^{-s}\}^{-1} \\ &= \prod_{p \nmid k_0 k} \{1 - \chi(p) p^{-s}\}^{-1} = f(s) L(s, \chi) \quad (\sigma > 1). \end{aligned}$$

Since, by Theorem 35, both $f_0(s) L(s, \chi_0)$ and $f(s) L(s, \chi)$ are regular for $\sigma > 0$, we deduce that

$$f_0(s) L(s, \chi_0) = f(s) L(s, \chi) \quad (\sigma > 0),$$

and the result follows since the only possible zeros of $f_0(s)$ and $f(s)$ are on the imaginary axis.

THEOREM 46. *Let* $k \geqslant 8$, *let* χ *be a non-principal character* (mod k), *and let* $s > 1 - 1/\log k$. *Then* $|L'(s, \chi)| \leqslant 6 \log^2 k$.

Proof. By (49),

$$L'(s, \chi) = -A - B,$$

where $A = \sum_{n=1}^{k} \chi(n) n^{-s} \log n, \quad B = \sum_{n=k+1}^{\infty} \chi(n) n^{-s} \log n.$

Now

$$\begin{aligned} |A| &\leqslant \sum_{n=1}^{k} n^{-1+1/\log k} \log n \leqslant k^{1/\log k} \log k \sum_{n=1}^{k} n^{-1} \\ &\leqslant e \log k (1 + \log k) < 5 \log^2 k. \end{aligned}$$

In order to obtain an inequality for $|B|$, define a_n as in the proof of Theorem 33. Then, as was shown there, $a_k = 0$ and

$$|a_n| \leqslant \tfrac{1}{2} k \quad (n = 0, 1, 2, \ldots).$$

Hence $$B = \sum_{n=k+1}^{\infty} (a_n - a_{n-1}) n^{-s} \log n$$
$$= \sum_{n=k+1}^{\infty} a_n\{n^{-s} \log n - (n+1)^{-s} \log (n+1)\}$$

and, since the hypotheses imply that $n^{-s}\log n$ is a decreasing function of n for $n \geqslant k$,

$$|B| \leqslant \tfrac{1}{2}k \sum_{n=k+1}^{\infty} \{n^{-s} \log n - (n+1)^{-s} \log (n+1)\}$$
$$= \tfrac{1}{2}k(k+1)^{-s} \log (k+1) < \tfrac{1}{2}k^{1-s} \log k < \tfrac{1}{2}e \log k < \log^2 k.$$

The result now follows.

THEOREM 47. *Let χ_0 and χ be non-equivalent non-principal real characters* (mod k_0) *and* (mod k) *respectively, and let*

$$g(s) = \zeta(s) L(s, \chi_0) L(s, \chi) L(s, \chi_0\chi),$$
$$\rho = L(1, \chi_0) L(1, \chi) L(1, \chi_0\chi),$$

and $$\tfrac{9}{10} < a < 1.$$

Then $$g(a) > \tfrac{1}{2} - \frac{6\rho}{1-a}(k_0 k)^{8(1-a)}.$$

Proof. By Theorem 43,

$$g(s) = \sum_{n=1}^{\infty} f(n) n^{-s} \quad (\sigma > 1),$$

where $f(n)$ satisfies (110) and (114). Hence $g(2) \geqslant 1$ and

$$(-1)^m g^{(m)}(2) = \sum_{n=1}^{\infty} f(n) n^{-2} \log^m n \geqslant 0 \quad (m = 1, 2, \ldots),$$

and hence, by Taylor's theorem for functions of a complex variable,

$$g(s) = \sum_{m=0}^{\infty} a_m (2-s)^m \quad (|s-2| < 1),$$

where $$a_0 \geqslant 1, \quad a_m \geqslant 0 \quad (m = 1, 2, \ldots). \tag{116}$$

It follows that

$$g(s)-\frac{\rho}{s-1}=\sum_{m=0}^{\infty}(a_m-\rho)(2-s)^m \tag{117}$$

for $|s-2|<1$. Now, by Theorems 1 and 35, $g(s)$ is regular for $\sigma>0$, except at $s=1$, where it has a simple pole with residue ρ. Hence $g(s)-\rho/(s-1)$ is regular for $\sigma>0$, and (117) holds not only for $|s-2|<1$, but for $|s-2|<2$. From this and Cauchy's inequality it follows that, if $|g(s)-\rho/(s-1)|\leqslant M$ $(|s-2|=\frac{3}{2})$, then

$$|a_m-\rho|\leqslant M(\tfrac{2}{3})^m \quad (m=0,1,2,\ldots).$$

Now let $|s-2|=\frac{3}{2}$. Then $|1/(s-1)|\leqslant 2$ and, by an argument like that preceding (23), $|s|/\sigma\leqslant 4/\sqrt{7}$. Hence, by (9), $|\zeta(s)|\leqslant 2+4/\sqrt{7}<10/\sqrt{7}$, and, since χ_0, χ, and $\chi_0\chi$ are non-principal characters (mod k_0), (mod k), and (mod k_0k) respectively, it follows from Theorem 35 that $|L(s,\chi_0)|\leqslant(4/\sqrt{7})k_0$, $|L(s,\chi)|\leqslant(4/\sqrt{7})k$, and $|L(s,\chi_0\chi)|\leqslant(4/\sqrt{7})k_0k$, and from (65) that $|L(1,\chi_0)|\leqslant k_0$, $|L(1,\chi)|\leqslant k$, and $|L(1,\chi_0\chi)|\leqslant k_0k$. Thus

$$|g(s)|\leqslant\frac{10}{\sqrt{7}}\left(\frac{4}{\sqrt{7}}\right)^3k_0^2k^2=\frac{640}{49}k_0^2k^2<14k_0^2k^2 \quad (|s-2|=\tfrac{3}{2})$$

and $|\rho|\leqslant k_0^2k^2$, so that

$$|g(s)-\rho/(s-1)|<16k_0^2k^2 \quad (|s-2|=\tfrac{3}{2}),$$

which means that we may take $M=16k_0^2k^2$ and obtain

$$|a_m-\rho|\leqslant 16k_0^2k^2(\tfrac{2}{3})^m \quad (m=0,1,2,\ldots). \tag{118}$$

Now, by (117), for any positive integer m_0,

$$g(a)=-\rho/(1-a)+A+B,$$

where

$$A=\sum_{m=0}^{m_0-1}(a_m-\rho)(2-a)^m, \quad B=\sum_{m=m_0}^{\infty}(a_m-\rho)(2-a)^m,$$

so that, by (116),

$$A\geqslant 1-\rho\sum_{m=0}^{m_0-1}(2-a)^m=1-\rho\frac{(2-a)^{m_0}-1}{1-a},$$

and, by (118) (since $a > \frac{9}{10}$),

$$B \geqslant -16k_0^2 k^2 \sum_{m=m_0}^{\infty} (\tfrac{2}{3})^m (\tfrac{11}{10})^m = -60k_0^2 k^2 (\tfrac{11}{15})^{m_0}.$$

Thus
$$g(a) \geqslant 1 - \frac{\rho(2-a)^{m_0}}{1-a} - 60k_0^2 k^2 (\tfrac{11}{15})^{m_0}.$$

Taking $m_0 = [\log(120k_0^2 k^2)/(\log 15 - \log 11)] + 1$, and noting that then $60k_0^2 k^2 (\frac{11}{15})^{m_0} < \frac{1}{2}$ and $m_0 < 8\log(k_0 k) + 17$, so that

$$(2-a)^{m_0} < (\tfrac{11}{10})^{17} (k_0 k)^{8\log(2-a)} < 6(k_0 k)^{8(1-a)},$$

we obtain the result stated.

THEOREM 48. *Corresponding to any ϵ, there is a positive number A such that, if $k > A$ and χ is a real non-principal character* (mod k), *then $L(s, \chi)$ has no zeros on the real axis between $1 - k^{-\epsilon}$ and* 1.

Proof. We may obviously restrict ourselves to the case $\epsilon < 1$. The result is trivial, viz. with $A = (\epsilon/16)^{-1/\epsilon}$, if no L function formed with a real non-principal character has a zero on the real axis between $1 - \epsilon/16$ and 1. Suppose, therefore, that there are a number k_0, a real non-principal character χ_0 (mod k_0), and a number a such that

$$1 - \epsilon/16 < a < 1 \tag{119}$$

and
$$L(a, \chi_0) = 0. \tag{120}$$

Then, since $L(s, \chi_0)$ cannot have infinitely many zeros between a and 1, there is a number a_1 such that $a \leqslant a_1 < 1$ and

$$L(s, \chi_0) \neq 0 \quad (a_1 < s < 1). \tag{121}$$

Also there is a number A_1 such that

$$\log k < k^{\epsilon} \quad (k > A_1) \tag{122}$$

and

$$\frac{72}{1-a} k_0^{8(1-a)} L(1, \chi_0) \{2 + \log(k_0 k)\} \log^2 k < k^{\epsilon/2} \quad (k > A_1). \tag{123}$$

We shall prove the result with

$$A = \max\{A_1, 8, (1-a_1)^{-1/\epsilon}\}.$$

Let $k > A$, and let χ be a real non-principal character (mod k). Then we have to prove that

$$L(s,\chi) \neq 0 \quad (1-k^{-\epsilon} < s < 1). \tag{124}$$

Suppose, first, that χ is equivalent to χ_0. Then (124) follows from (121) and Theorem 45 since $k > (1-a_1)^{-1/\epsilon}$. Now suppose that χ is not equivalent to χ_0. Then, by (120) and Theorem 47,

$$0 > \frac{1}{2} - \frac{6}{1-a} L(1,\chi_0)\, L(1,\chi)\, L(1,\chi_0\chi)\, (k_0 k)^{8(1-a)},$$

and it easily follows from (66) that $L(1,\chi_0) \geqslant 0$ and $L(1,\chi) \geqslant 0$. Hence, by Theorem 36,

$$\frac{12}{1-a} L(1,\chi_0)\, L(1,\chi)\,\{2+\log(k_0 k)\}\,(k_0 k)^{8(1-a)} > 1,$$

and hence, by (123),

$$L(1,\chi)\, k^{8(1-a)+\epsilon/2} > 6 \log^2 k.$$

From this and (119) it follows that

$$L(1,\chi) > 6k^{-\epsilon} \log^2 k. \tag{125}$$

Now suppose, if possible, that (124) is false. Then there is a number a_2 such that

$$1-k^{-\epsilon} < a_2 < 1 \tag{126}$$

and $$L(a_2,\chi) = 0.$$

From this and the mean value theorem of the differential calculus it follows that there is a number $a_3 > 1-k^{-\epsilon}$ such that

$$L(1,\chi) = (1-a_2)\, L'(a_3,\chi). \tag{127}$$

Now, by (122), $a_3 > 1-1/\log k$, and hence, by Theorem 46, $L'(a_3,\chi) \leqslant 6\log^2 k$. From this and (127) and (126) we obtain $L(1,\chi) < 6k^{-\epsilon}\log^2 k$, which contradicts (125). This completes the proof.

2·13. In the remainder of this chapter, u denotes a positive number, as large as we please, and $A_1, A_2, \ldots$ are suitable (sufficiently large) positive numbers depending only on u.

THEOREM 49. *Let*

$$b > A_1, \tag{128}$$

$$|t| \leqslant 2b, \tag{129}$$

$$k \leqslant \log^{3u} b, \tag{130}$$

$$\sigma > 1 - 1/\{8000 \log (2b)\}, \tag{131}$$

and let χ *be a non-principal character* (mod k). *Then* $L(s, \chi) \neq 0$.

Proof. Suppose first that χ is not real. Then, if A_1 is large enough, it follows from (128), (129), (130), and (69) that $2b \geqslant \tau$. Hence the inequality (131) implies that $\sigma > 1 - 1/(4000 \log \tau)$, and the result follows from Theorem 39.

Next suppose that χ is real and $t \neq 0$. Then the result follows similarly from Theorem 40.

Lastly, suppose that χ is real and $t = 0$, which means that s is also real. Then, in the cases $s > 1$ and $s = 1$, the result follows from (67) and Theorem 44 respectively. Thus all that remains is to consider the case

$$1 - 1/\{8000 \log (2b)\} < s < 1. \tag{132}$$

Now, by Theorem 48 with $\epsilon = 1/(3u)$, there is a number A, depending on u only, such that, if $k > A$ and

$$1 - k^{-1/(3u)} < s < 1, \tag{133}$$

then $L(s, \chi) \neq 0$, and it is clear that (130) and (132) imply (133). Finally there are only a finite number of L functions, formed with characters (mod k), where $k \leqslant A$. Hence there is a number $a < 1$ such that none of these functions has a zero on the real axis between a and 1. In other words, if $k \leqslant A$ and

$$a < s < 1, \tag{134}$$

then $L(s, \chi) \neq 0$, and it is clear that (128) and (132) imply (134) if A_1 is large enough. This completes the proof.

THEOREM 50. *Let* (128) *and* (130) *hold, and let* χ *be a non-principal character* (mod k). *Then* $L'(s,\chi)/L(s,\chi)$ *is regular in the set of points given by* (129) *and* (131).

This follows from Theorems 35 and 49.

THEOREM 51. *Let* (128) *and* (130) *hold, let* χ *be a non-principal character* (mod k), *and let*

$$|t| \leqslant b \tag{135}$$

and

$$1 - 1/(20000 \log b) \leqslant \sigma < 2. \tag{136}$$

Then

$$|L'(s,\chi)/L(s,\chi)| \leqslant C_7 \log^3 b.$$

Proof. Let $r_1 = 1 + 1/(10000 \log b)$, $r_2 = 1 + 1/(9000 \log b)$, and

$$g(z) = \int_{2+it}^{2+it+z} \frac{L'(w,\chi)}{L(w,\chi)}\, dw.$$

Then, by Theorem 50, $g(z)$ is regular for $|z| < r_2$. Also

$$e^{g(z)} = L(2+it+z,\chi)/L(2+it,\chi).$$

Now, by Theorem 35, (128), and (130),

$$|L(2+it+z,\chi)| \leqslant k(4+b)/(2-r_1) < \tfrac{1}{2}b^2 \quad (|z| \leqslant r_1),$$

and, by (66) (cf. the proof of Theorem 14),

$$|L(2+it,\chi)|^{-1} = \prod_p |1-\chi(p)p^{-2-it}| \leqslant \prod_p (1+p^{-2}) < 2.$$

Hence $e^{\mathbf{R}g(z)} < b^2 (|z| \leqslant r_1)$, and the hypotheses of Theorem 13 are satisfied with $M = 2 \log b$. It follows that

$$|L'(s,\chi)/L(s,\chi)| = |g'(s-2-it)| = |g'(\sigma-2)| \leqslant 2Mr_1(r_1+\sigma-2)^{-2}$$
$$\leqslant 2Mr_1(20000 \log b)^2 < 2 \times 10^9 \log^3 b,$$

which proves the theorem.

2·14. The theory of the L functions can now be applied to the problem of the distribution of primes. We put

$$\psi(m:\chi) = \sum_{n=1}^{m} \Lambda(n)\chi(n) \tag{137}$$

and

$$\vartheta(m;\chi) = \sum_{p \leqslant m} \chi(p) \log p. \tag{138}$$

Then, with $E(x)$ defined by (35),

$$\psi(m;\chi) = \sum_{n=1}^{\infty} E\left(\frac{m+\frac{1}{2}}{n}\right) \Lambda(n)\chi(n) \quad (m = 1, 2, \ldots). \tag{139}$$

THEOREM 52. *Let $m \geqslant 3$,*

$$k \leqslant \log^u m, \tag{140}$$

and let χ be a non-principal character (mod k). *Then*

$$|\psi(m;\chi)| \leqslant A_2 m \exp\left(-\frac{\sqrt{\log m}}{200}\right).$$

Proof. Suppose, first, that

$$m > A_3, \tag{141}$$

and let
$$a = 1 + 1/\log(m+\tfrac{1}{2})$$

and
$$b = \exp\frac{\sqrt{\log(m+\frac{1}{2})}}{160}.$$

Then, by (139), (68), and Theorems 15 and 17,

$$\left|\psi(m;\chi) + \frac{1}{2\pi i}\int_{a-ib}^{a+ib} \frac{(m+\frac{1}{2})^s}{s}\frac{L'(s,\chi)}{L(s,\chi)}\,ds\right|$$
$$= \left|\sum_{n=1}^{\infty} \Lambda(n)\chi(n)\left\{E\left(\frac{m+\frac{1}{2}}{n}\right) - \frac{1}{2\pi i}\int_{a-ib}^{a+ib} \frac{1}{s}\left(\frac{m+\frac{1}{2}}{n}\right)^s ds\right\}\right|$$
$$\leqslant \frac{1}{\pi b}\sum_{n=1}^{\infty} \log n \left(\frac{m+\frac{1}{2}}{n}\right)^a \left|\log\frac{m+\frac{1}{2}}{n}\right|^{-1} < \frac{22}{\pi b} m(3+\log m)^2.$$

It follows that

$$|\psi(m;\chi)| \leqslant \left|\int_{a-ib}^{a+ib} \frac{(m+\frac{1}{2})^s}{s}\frac{L'(s,\chi)}{L(s,\chi)}\,ds\right| + C_8 m \exp\left(-\frac{\sqrt{\log m}}{200}\right). \tag{142}$$

Now, if A_3 is large enough, (141) and (140) imply (128) and (130). Hence, by Theorems 50 and 51, $L'(s,\chi)/L(s,\chi)$ is regular, and $|L'(s,\chi)/L(s,\chi)| \leqslant C_7 \log^3 b$, within and on the rectangle with corners at $a \pm ib$ and $a' \pm ib$, where

$$a' = 1 - 1/(20000 \log b).$$

From this and Cauchy's theorem it follows that, if C denotes the broken line $\{a-ib, a'-ib, a'+ib, a+ib\}$, then

$$\left| \int_{a-ib}^{a+ib} \frac{(m+\frac{1}{2})^s}{s} \frac{L'(s,\chi)}{L(s,\chi)} ds \right| = \left| \int_C \frac{(m+\frac{1}{2})^s}{s} \frac{L'(s,\chi)}{L(s,\chi)} ds \right|$$

$$\leqslant 2C_7 \log^3 b \left\{ \int_0^b \frac{(m+\frac{1}{2})^{a'}}{|a'+it|} dt + \int_{a'}^a \frac{(m+\frac{1}{2})^\sigma}{b} d\sigma \right\}$$

$$\leqslant 2C_7 \log^3 b \{ (m+\tfrac{1}{2})^{a'} (2 + \log b) + (m+\tfrac{1}{2})^a / b \}$$

$$\leqslant C_8' m \log^4 b \left\{ \exp\left(-\frac{\sqrt{\log (m+\frac{1}{2})}}{125} \right) + \exp\left(-\frac{\sqrt{\log (m+\frac{1}{2})}}{160} \right) \right\}$$

$$\leqslant C_9 m \exp\left(-\frac{\sqrt{\log m}}{200} \right).$$

Hence, by (142),

$$|\psi(m;\chi)| \leqslant (C_8 + C_9) m \exp\left(-\frac{\sqrt{\log m}}{200} \right).$$

Now suppose that $m \leqslant A_3$. Then we use the trivial inequality $|\psi(m;\chi)| \leqslant m \log m$, and obtain

$$|\psi(m;\chi)| \leqslant m \log A_3 \leqslant \log A_3 \exp \frac{\sqrt{\log A_3}}{200} m \exp\left(-\frac{\sqrt{\log m}}{200} \right).$$

Thus the result holds with

$$A_2 = \max\left(C_8 + C_9, \log A_3 \exp \frac{\sqrt{\log A_3}}{200} \right).$$

2·15. Using the trivial inequality (cf. (38))

$$|\psi(m;\chi) - \vartheta(m;\chi)| \leqslant \sqrt{m} \log^2 m \quad (m \geqslant 1),$$

we deduce that, on the hypotheses of Theorem 52,

$$|\vartheta(m;\chi)| \leqslant A_4 m \exp\left(-\frac{\sqrt{\log m}}{200} \right).$$

Replacing u by $2u$ and at the same time A_4 by A_5, which is obviously permissible, we obtain

THEOREM 53. *Let* $m \geqslant 3$, $k \leqslant \log^{2u} m$, *and let* χ *be a non-principal character* (mod k). *Then*

$$|\vartheta(m;\chi)| \leqslant A_5 m \exp\left(-\frac{\sqrt{\log m}}{200}\right).$$

THEOREM 54. *Let* $n \geqslant 3$, $k \leqslant \log^{u} n$, *and let* χ *be a non-principal character* (mod k). *Then*

$$|\pi(n;\chi)| \leqslant A_6 n \exp\left(-\frac{\sqrt{\log n}}{200}\right).$$

Proof. We begin by showing that

$$|\vartheta(m;\chi)| \leqslant A_7 n \exp\left(-\frac{\sqrt{\log n}}{200}\right) \quad (1 \leqslant m \leqslant n). \tag{143}$$

Suppose, first, that $\exp\sqrt{\log n} \leqslant m \leqslant n$. Then $\log^2 m \geqslant \log n$, so that $k \leqslant \log^{2u} m$, which implies that $m \geqslant 3$. Hence, by Theorem 53,

$$|\vartheta(m;\chi)| \leqslant A_5 m \exp\left(-\frac{\sqrt{\log m}}{200}\right) \leqslant A_5 n \exp\left(-\frac{\sqrt{\log n}}{200}\right).$$

Now let $1 \leqslant m \leqslant \exp\sqrt{\log n}$. Then

$$|\vartheta(m;\chi)| \leqslant m \log m < m^2 \leqslant \exp(2\sqrt{\log n}) \leqslant C_{10} n \exp\left(-\frac{\sqrt{\log n}}{200}\right).$$

Thus (143) holds with $A_7 = \max(A_5, C_{10})$.

Now, by (46) and (138),

$$\pi(n;\chi) = \sum_{m=2}^{n} \frac{\vartheta(m;\chi) - \vartheta(m-1;\chi)}{\log m}$$

$$= \sum_{m=2}^{n-1} \vartheta(m;\chi)\left(\frac{1}{\log m} - \frac{1}{\log(m+1)}\right) + \frac{\vartheta(n;\chi)}{\log n}.$$

Hence, by (143),

$$|\pi(n;\chi)| \leqslant A_7 n \exp\left(-\frac{\sqrt{\log n}}{200}\right)\left\{\sum_{m=2}^{n-1}\left(\frac{1}{\log m}-\frac{1}{\log(m+1)}\right)+\frac{1}{\log n}\right\}$$

$$= \frac{A_7}{\log 2} n \exp\left(-\frac{\sqrt{\log n}}{200}\right),$$

which proves the theorem.

THEOREM 55. *Let* $m \geqslant 3$, $k \leqslant \log^u m$, *and* $(k,l)=1$. *Then*

$$\left|\pi(m;k,l)-\frac{\operatorname{ls} m}{\phi(k)}\right| \leqslant A_8 m \exp\left(-\frac{\sqrt{\log m}}{200}\right).$$

Proof. Denoting, as in § 2·2, the principal character (mod k) by χ_1 and the remaining characters (mod k) (if any) by $\chi_2, \chi_3, \ldots, \chi_{\phi(k)}$, we have, by (47), (48), and Theorem 54,

$$\left|\pi(m;k,l)-\frac{\pi(m)}{\phi(k)}\right| \leqslant 1+\frac{1}{\phi(k)}\sum_{h=2}^{\phi(k)}|\pi(m;\chi_h)|$$

$$\leqslant 1 + A_6 m \exp\left(-\frac{\sqrt{\log m}}{200}\right).$$

From this and Theorem 19 we obtain the result stated.

CHAPTER 3

THE REPRESENTATIONS OF AN ODD NUMBER AS A SUM OF THREE PRIMES

3·1. It was first proved by Vinogradoff in 1937 that every odd $n > C_{11}$ can be represented in the form $p_1 + p_2 + p_3$, but Hardy and Littlewood showed in 1923 that this is so if there is a number $\theta < \frac{3}{4}$ such that none of Dirichlet's L functions has zeros in the half-plane $\sigma > \theta$.

Again we begin with an outline of the proof. We ask ourselves not only whether the number n can be represented in the form stated, but also in how many ways. Let $r(n)$ denote the number of ways, i.e. the number of solutions of the equation $p_1 + p_2 + p_3 = n$ in primes p_1, p_2, p_3. Then we have to prove that

$$r(n) > 0 \quad (2 \nmid n, n > C_{11}). \tag{144}$$

We do this by obtaining an asymptotic formula for $r(n)$.

From now on we assume that

$$n > C_{11} \tag{145}$$

throughout. Putting

$$f(x, v) = \sum_{p \leqslant v} e(px) \quad (v \geqslant 0), \tag{146}$$

we have

$$f^3(x, n) = \sum_{p_1 \leqslant n} \sum_{p_2 \leqslant n} \sum_{p_3 \leqslant n} e\{(p_1 + p_2 + p_3) x\} = \sum_{m=6}^{3n} r(m, n)\, e(mx),$$

where $r(m, n)$ is the number of representations of m as a sum of three primes, none of which exceeds n. In particular $r(n, n) = r(n)$. Hence

$$r(n) = \int_{x_0}^{x_0+1} f^3(x, n)\, e(-nx)\, dx \tag{147}$$

for any x_0. For convenience, we take

$$x_0 = n^{-1} \log^{15} n. \tag{148}$$

We shall find that $f(x,n)$ is comparatively small unless x is near a rational number with a small denominator, in which case there is a suitable approximating function to $f(x,n)$, though different approximating functions correspond to different rational numbers. This suggests that we divide the interval (x_0, x_0+1) (the interval of integration in (147)) into two parts E_1 and E_2, say, where E_1 is the set of those numbers in the above interval which are not near rational numbers with small denominators, and E_2 consists of intervals about such rational numbers. Here the precise meaning of 'near' and 'small' is unimportant. For convenience, we interpret 'near' as meaning 'at a distance not exceeding x_0' and 'small' as 'not exceeding $\log^{15} n$'. Thus, by (147),

$$r(n) = \int_{E_1} f^3(x,n)\,e(-nx)\,dx + \sum_{q \leqslant \log^{15} n} \sum_{\substack{0<h\leqslant q \\ (h,q)=1}} J(h,q), \quad (149)$$

where
$$J(h,q) = \int_{h/q-x_0}^{h/q+x_0} f^3(x,n)\,e(-nx)\,dx. \quad (150)$$

We note that the fractions h/q associated with the pairs of numbers h,q occurring in (149) are the fractions with 'small' denominators in the interval (x_0, x_0+1), and that, by (145) and (148), the corresponding intervals $(h/q-x_0, h/q+x_0)$ are non-overlapping.

3·2. The function $f(x,v)$, defined by (146), like all functions of the form

$$\sum_{0<m\leqslant v} a_m e(mx),$$

satisfies the identity

$$f(x_1+x_2, v) = e(vx_2) f(x_1, v) - 2\pi i x_2 \int_0^v e(ux_2) f(x_1, u)\,du. \quad (151)$$

To prove this, we note that

$$e(vx_2) - e(px_2) = \int_p^v 2\pi i x_2 e(ux_2)\,du.$$

Hence, by (146),

$$\begin{aligned} f(x_1+x_2, v) &= \sum_{p \leqslant v} e(px_1)\, e(px_2) \\ &= \sum_{p \leqslant v} e(px_1) \left\{ e(vx_2) - 2\pi i x_2 \int_p^v e(ux_2)\, du \right\} \\ &= e(vx_2) \sum_{p \leqslant v} e(px_1) - 2\pi i x_2 \int_0^v e(ux_2) \sum_{p \leqslant u} e(px_1)\, du \\ &= e(vx_2) f(x_1, v) - 2\pi i x_2 \int_0^v e(ux_2) f(x_1, u)\, du. \end{aligned}$$

3·3. Our next main aim is to prove that

$$|f(x,n)| \leqslant C_{12} n \log^{-3} n \quad (x \in E_1). \tag{152}$$

We shall deduce this from (151) and the following theorem, whose proof, though elementary, is complicated.

THEOREM 56. *Let*

$$n \log^{-3} n < v \leqslant n, \tag{153}$$

$$\log^{15} n < q \leqslant n \log^{-15} n, \tag{154}$$

and

$$(h, q) = 1. \tag{155}$$

Then

$$|f(h/q, v)| \leqslant n \log^{-3} n. \tag{156}$$

Note. In the proof that follows, the letters j, m, ν, and ξ (as well as k and q) denote positive integers.

Proof. Let

$$a_1 = \prod_{p \leqslant \sqrt{n}} p. \tag{157}$$

Then the numbers $m \leqslant v$ such that $(m, a_1) = 1$ are the number 1 and the primes between $\sqrt{n}$ (exclusive) and v (inclusive). Hence, putting

$$b_1 = \sum_{\substack{m \leqslant v \\ (m, a_1) = 1}} e(mh/q), \tag{158}$$

we have, by (146),

$$|f(h/q, v) - b_1| \leqslant \sqrt{n}. \tag{159}$$

Now, by H.-W., Theorem 264,

$$\sum_{\substack{j,k\\ j\mid a_1,\, jk=m}} \mu(j) = \sum_{j\mid(m,a_1)} \mu(j) = \begin{cases} 1 & ((m,a_1)=1)\\ 0 & ((m,a_1)>1),\end{cases}$$

where μ denotes Möbius's function. Hence, by (158),

$$b_1 = \sum_{m\leqslant v} e\left(\frac{mh}{q}\right) \sum_{\substack{j,k\\ j\mid a_1,\, jk=m}} \mu(j) = \sum_{\substack{j,k\\ j\mid a_1,\, jk\leqslant v}} \mu(j) e\left(\frac{hjk}{q}\right) = b_2 + b_3, \quad (160)$$

where
$$b_2 = \sum_{\substack{j\mid a_1\\ j\leqslant v\log^{-4} n}} \mu(j) \sum_{k\leqslant v/j} e\left(\frac{hjk}{q}\right) \quad (161)$$

and
$$b_3 = \sum_{k<\log^4 n} \sum_{\substack{j\mid a_1\\ v\log^{-4} n<j\leqslant v/k}} \mu(j) e\left(\frac{hjk}{q}\right). \quad (162)$$

We first deal with b_2. By (161),

$$|b_2| \leqslant \sum_{j\leqslant v\log^{-4} n} \left| \sum_{k\leqslant v/j} e\left(\frac{hjk}{q}\right)\right| = \sum_{\substack{l\\ -\frac{1}{2}q<l\leqslant\frac{1}{2}q}} b_4(l), \quad (163)$$

where
$$b_4(l) = \sum_{\substack{j\leqslant v\log^{-4} n\\ hj\equiv l \pmod q}} \left| \sum_{k\leqslant v/j} e\left(\frac{lk}{q}\right)\right|. \quad (164)$$

From this and (155) and (153) it follows that

$$b_4(0) = \sum_{\substack{j\leqslant v\log^{-4} n\\ q\mid j}} \left[\frac{v}{j}\right] = \sum_{m\leqslant q^{-1}v\log^{-4} n} \left[\frac{v}{qm}\right] \leqslant \frac{n}{q} \sum_{m\leqslant n} \frac{1}{m},$$

and hence, by (154), $b_4(0) \leqslant 2n\log^{-14} n$. (165)

Now let $0<|l|\leqslant\frac{1}{2}q$. Then

$$\left|\sum_{k\leqslant v/j} e\left(\frac{lk}{q}\right)\right| = \left|\frac{e(l[v/j]/q)-1}{e(l/q)-1}\right| \leqslant \frac{2}{|e(l/q)-1|} = \frac{1}{|\sin(\pi l/q)|} \leqslant \frac{q}{2|l|},$$

and hence, by (164) and (155),

$$b_4(l) \leqslant \frac{q}{2|l|} \sum_{\substack{j \leqslant v\log^{-6} n \\ hj \equiv l \pmod q}} 1 \leqslant \frac{1}{2|l|}(v\log^{-5} n + q),$$

so that, by (153) and (154),

$$\sum_{\substack{l \neq 0 \\ -\frac{1}{2}q < l \leqslant \frac{1}{2}q}} b_4(l) \leqslant (v\log^{-5} n + q) \sum_{0 < l \leqslant \frac{1}{2}q} l^{-1} \leqslant 2n\log^{-4} n.$$

From this and (163) and (165) it follows that

$$|b_2| \leqslant 3n\log^{-4} n. \tag{166}$$

Turning to b_3, we put

$$b_5(k) = \sum_{\substack{j \mid a_1 \\ v\log^{-6} n < j \leqslant v/k}} \mu(j)\, e\left(\frac{hjk}{q}\right). \tag{167}$$

Then, by (162),

$$b_3 = \sum_{k < \log^6 n} b_5(k). \tag{168}$$

Now let

$$a_2 = \prod_{p \leqslant \log^{15} n} p. \tag{169}$$

Then, by (167), (157), and (145),

$$b_5(k) = b_6 + b_7, \tag{170}$$

where

$$b_6 = \sum_{\substack{j \mid a_2 \\ v\log^{-6} n < j \leqslant v/k}} \mu(j)\, e\left(\frac{hjk}{q}\right) \tag{171}$$

and

$$b_7 = \sum_{\substack{j \mid a_1,\ j \nmid a_2 \\ v\log^{-6} n < j \leqslant v/k}} \mu(j)\, e\left(\frac{hjk}{q}\right). \tag{172}$$

Obviously, by (171) and (153),

$$|b_6| \leqslant \sum_{\substack{j \mid a_2 \\ n\log^{-6} n < j \leqslant n}} 1. \tag{173}$$

Now let
$$\omega(m) = \sum_{p|m} 1, \quad d(m) = \sum_{\nu|m} 1 \tag{174}$$

(H.-W., §§ 22·12 and 16·7). Then, for any $j \mid a_2$, by (169), $j \leqslant (\log^{15} n)^{\omega(j)}$ and $d(j) = 2^{\omega(j)}$, so that $d(j) \geqslant j^{(\log 2)/(15 \log \log n)}$. Hence, by (173),

$$|b_6| (n \log^{-8} n)^{(\log 2)/(15 \log \log n)} \leqslant \sum_{j \leqslant n} d(j) = \sum_{j \leqslant n} \sum_{\nu | j} 1 = \sum_{\nu \leqslant n} \sum_{\substack{j \leqslant n \\ \nu | j}} 1$$
$$= \sum_{\nu \leqslant n} \left[\frac{n}{\nu}\right] \leqslant n \sum_{\nu \leqslant n} \nu^{-1} < n(1 + \log n).$$

From this and (145) it follows that

$$|b_6| \leqslant n \log^{-100} n. \tag{175}$$

In order to investigate b_7, we put

$$b_8(x) = \sum_{\substack{j | a_1, j \nmid a_2 \\ j \leqslant x}} \mu(j)\, e\left(\frac{hjk}{q}\right). \tag{176}$$

Then, by (172),

$$b_7 = b_8(v/k) - b_8(v \log^{-5} n) \quad (k < \log^5 n). \tag{177}$$

In dealing with $b_8(x)$, we may assume that

$$0 < x \leqslant n/k. \tag{178}$$

If $j \mid a_1$, it follows from (157), (169), and (174) that the number of those prime divisors of j which are greater than $\log^{15} n$ is $\omega\{(j, a_1/a_2)\}$. If, moreover, $j \nmid a_2$ and $j \leqslant n$, it follows that $1 \leqslant \omega\{(j, a_1/a_2)\} \leqslant \log n$. Hence, by (176) and (178),

$$b_8(x) = \sum_{m \leqslant \log n} b_9(m), \tag{179}$$

where
$$b_9(m) = \sum_{\substack{j | a_1, j \leqslant x \\ \omega\{(j, a_1/a_2)\} = m}} \mu(j)\, e\left(\frac{hjk}{q}\right). \tag{180}$$

From this and (174) it follows that

$$mb_9(m) = \sum_{\substack{j \mid a_1,\, j<x \\ \omega\{(j,\, a_1/a_2)\}=m}} \mu(j)\, e\!\left(\frac{hjk}{q}\right) \sum_{\substack{p,\, \nu \\ p\nu=j,\, p \mid (a_1/a_2)}} 1$$

$$= \sum_{\substack{p,\, \nu \\ p\nu \mid a_1,\, p\nu<x \\ \omega\{(p\nu,\, a_1/a_2)\}=m,\, p \mid (a_1/a_2)}} \mu(p\nu)\, e\!\left(\frac{hp\nu k}{q}\right)$$

$$= -\sum_{p \mid (a_1/a_2)} \sum_{\substack{\nu \mid (a_1/p),\, \nu<x/p \\ \omega\{(\nu,\, a_1/a_2)\}=m-1}} \mu(\nu)\, e\!\left(\frac{hp\nu k}{q}\right).$$

Note that, in virtue of (157) and (169), the condition $p \mid (a_1/a_2)$ is equivalent to $\log^{15} n < p \leqslant \sqrt{n}$, and that, if this is satisfied, then the condition $\nu \mid (a_1/p)$ is equivalent to the simultaneous conditions $\nu \mid a_1$ and $p \nmid \nu$. Hence

$$mb_9(m) = -b_{10} + b_{11}, \tag{181}$$

where

$$b_{10} = \sum_{\log^{15} n < p \leqslant \sqrt{n}} \sum_{\substack{\nu \mid a_1,\, \nu \leqslant x/p \\ \omega\{(\nu,\, a_1/a_2)\}=m-1}} \mu(\nu)\, e\!\left(\frac{hp\nu k}{q}\right) \tag{182}$$

and

$$b_{11} = \sum_{\log^{15} n < p \leqslant \sqrt{n}} \sum_{\substack{\nu \mid a_1,\, p \mid \nu,\, \nu \leqslant x/p \\ \omega\{(\nu,\, a_1/a_2)\}=m-1}} \mu(\nu)\, e\!\left(\frac{hp\nu k}{q}\right),$$

so that

$$|b_{11}| \leqslant \sum_{p>\log^{15} n} \sum_{\nu \leqslant x/p,\, p \mid \nu} 1 = \sum_{p>\log^{15} n} [xp^{-2}] < 2x \log^{-15} n,$$

and hence, by (178),

$$|b_{11}| < 2nk^{-1} \log^{-15} n. \tag{183}$$

It remains to investigate b_{10}. For brevity, we put $q/(q,k) = \alpha$ and $hk/(q,k) = \beta$. Then, by (155),

$$(\alpha, \beta) = 1. \tag{184}$$

Also, since $q/k \leqslant \alpha \leqslant q$, it follows from (154) that

$$k^{-1} \log^{15} n < \alpha \leqslant n \log^{-15} n. \tag{185}$$

We now introduce the auxiliary function $g(\nu)$, defined as $\mu(\nu)$ or 0 according as the conditions $\nu \mid a_1$ and $\omega\{(\nu, a_1/a_2)\} = m-1$ are or are not both satisfied. Then

$$|g(\nu)| \leqslant 1, \tag{186}$$

and it follows from (182) that

$$b_{10} = \sum_{\log^{15} n < p \leqslant \sqrt{n}} \sum_{\nu \leqslant x/p} g(\nu)\, e\left(\frac{\beta p \nu}{\alpha}\right). \tag{187}$$

Let λ be the integer for which

$$2^{\lambda-1} < \sqrt{n} \log^{-15} n \leqslant 2^{\lambda}. \tag{188}$$

Then, by (187),

$$|b_{10}| \leqslant \sum_{\log^{15} n < j \leqslant 2^{\lambda} \log^{15} n} \left| \sum_{\nu \leqslant x/j} g(\nu)\, e\left(\frac{\beta j \nu}{\alpha}\right) \right| = \sum_{\xi \leqslant \lambda} b_{12}(\xi), \tag{189}$$

where $$b_{12}(\xi) = \sum_{2^{\xi-1} \log^{15} n < j \leqslant 2^{\xi} \log^{15} n} \left| \sum_{\nu \leqslant x/j} g(\nu)\, e\left(\frac{\beta j \nu}{\alpha}\right) \right|.$$

From this and Cauchy's inequality we obtain

$$b_{12}^2(\xi) \leqslant 2^{\xi} \log^{15} n \sum_{2^{\xi-1} \log^{15} n < j \leqslant 2^{\xi} \log^{15} n} \left| \sum_{\nu \leqslant x/j} g(\nu)\, e\left(\frac{\beta j \nu}{\alpha}\right) \right|^2$$

$$= 2^{\xi} \log^{15} n \,.\, b_{13}, \tag{190}$$

where

$$b_{13} = \sum_{2^{\xi-1} \log^{15} n < j \leqslant 2^{\xi} \log^{15} n} \sum_{\nu \leqslant x/j} \sum_{\nu' \leqslant x/j} g(\nu) g(\nu')\, e\left(\frac{\beta j(\nu - \nu')}{\alpha}\right)$$

$$= \sum_{\nu < x_1} \sum_{\nu' < x_1} g(\nu) g(\nu') \sum_{x_2 < j \leqslant x_3} e\left(\frac{\beta j(\nu - \nu')}{\alpha}\right), \tag{191}$$

with $$x_1 = x 2^{1-\xi} \log^{-15} n, \tag{192}$$

$$x_2 = 2^{\xi-1} \log^{15} n, \tag{193}$$

and $$x_3 = \min(2^{\xi} \log^{15} n, x/\nu, x/\nu'). \tag{194}$$

By (191) and (186),

$$b_{13} \leqslant \sum_{\nu<x_1} \sum_{\nu'<x_1} \left| \sum_{x_1<j\leqslant x_2} e\left(\frac{\beta j(\nu-\nu')}{\alpha}\right) \right| = \sum_{-\frac{1}{2}\alpha<l\leqslant\frac{1}{2}\alpha} b_{14}(l), \quad (195)$$

where

$$b_{14}(l) = \sum_{\nu<x_1} \sum_{\substack{\nu'<x_1 \\ \beta(\nu-\nu')\equiv l \pmod{\alpha}}} \left| \sum_{x_1<j\leqslant x_2} e\left(\frac{lj}{\alpha}\right) \right|. \quad (196)$$

The splitting up has at last come to an end. Now, by (184),

$$\sum_{\substack{\nu'<x_1 \\ \beta(\nu-\nu')\equiv l \pmod{\alpha}}} 1 < \left(\frac{x_1}{\alpha}+1\right),$$

and hence

$$\sum_{\nu<x_1} \sum_{\substack{\nu'<x_1 \\ \beta(\nu-\nu')\equiv l \pmod{\alpha}}} 1 < x_1\left(\frac{x_1}{\alpha}+1\right). \quad (197)$$

From this and (196), (194), and (192) it follows that

$$b_{14}(0) < 2^{\xi} \log^{15} n \,.\, x_1\left(\frac{x_1}{\alpha}+1\right) = 2x\left(\frac{x_1}{\alpha}+1\right). \quad (198)$$

Also (cf. the argument immediately after (165))

$$\left| \sum_{x_1<j\leqslant x_2} e\left(\frac{lj}{\alpha}\right) \right| \leqslant \frac{\alpha}{2\,|\,l\,|} \quad (0<|\,l\,|\leqslant \tfrac{1}{2}\alpha),$$

and hence, by (196) and (197),

$$b_{14}(l) \leqslant (2\,|\,l\,|)^{-1} x_1(x_1+\alpha) \quad (0<|\,l\,|\leqslant \tfrac{1}{2}\alpha). \quad (199)$$

The remainder of the proof is straightforward. We trace our way back to $f(h/q, v)$. By (195), (198), and (199),

$$b_{13} \leqslant 2x\left(\frac{x_1}{\alpha}+1\right) + x_1(x_1+\alpha)(1+\log\alpha).$$

From this and (178), (185), and (192) it follows that

$$b_{13} \leqslant 2\frac{n}{k}(n2^{1-\xi}\log^{-30} n+1)$$
$$+\frac{n}{k}2^{1-\xi}\log^{-15} n\left(\frac{n}{k}2^{1-\xi}\log^{-15} n + n\log^{-15} n\right)\log n,$$

so that, by (190), (188), and (145),

$$b_{12}^2(\xi) \leqslant 2\frac{n}{k}(2n\log^{-15}n + 2^{\xi}\log^{15}n) + 2\frac{n}{k}\log^{-14}n\left(\frac{n}{k}2^{1-\xi}+n\right)$$
$$\leqslant 5n^2k^{-1}\log^{-14}n \quad (\xi \leqslant \lambda).$$

Hence, by (189) and (188),

$$|b_{10}| \leqslant 3\lambda nk^{-\frac{1}{2}}\log^{-7}n \leqslant 3nk^{-\frac{1}{2}}\log^{-6}n.$$

From this and (181) and (183) it follows that

$$m\,|b_9(m)| \leqslant 4nk^{-\frac{1}{2}}\log^{-6}n,$$

from this and (179) that

$$|b_8(x)| \leqslant 5nk^{-\frac{1}{2}}\log^{-6}n\log\log n,$$

from this and (177) that

$$|b_7| \leqslant 10nk^{-\frac{1}{2}}\log^{-6}n\log\log n \quad (k < \log^5 n),$$

from this and (170) and (175) that

$$|b_5(k)| \leqslant 11nk^{-\frac{1}{2}}\log^{-6}n\log\log n \quad (k < \log^5 n),$$

from this and (168) that

$$|b_3| \leqslant 22n\log^{-3\frac{1}{2}}n\log\log n < \tfrac{1}{3}n\log^{-3}n,$$

from this and (160) and (166) that

$$|b_1| \leqslant \tfrac{2}{3}n\log^{-3}n,$$

and from this and (159) that

$$|f(h/q, v)| \leqslant n\log^{-3}n,$$

which proves the theorem.

3·4. To prove (152), we use the following well-known elementary theorem (cf. H.-W., Theorem 36).

THEOREM 57. *Corresponding to any x and any $y \geqslant 1$, there are numbers h and q satisfying* (155) *such that $q \leqslant y$ and*

$$|qx - h| < 1/y.$$

Now suppose that $x \in E_1$. Then, by Theorem 57, there are numbers h and q satisfying (155) such that

$$q \leqslant n \log^{-15} n$$

and $$|qx - h| < n^{-1} \log^{15} n, \tag{200}$$

and we cannot have $q \leqslant \log^{15} n$, for this, together with (200) and (148), would imply that x is 'near' the rational number h/q with the 'small' denominator q, and therefore, by § 3·1, not in E_1. Hence (154) is satisfied, and it follows from Theorem 56 and the trivial inequality $|f(h/q, v)| \leqslant v$ that

$$|f(h/q, v)| \leqslant n \log^{-3} n \quad (0 \leqslant v \leqslant n). \tag{201}$$

Putting $y = x - h/q$, we have, by (200) and (154), $|y| < n^{-1}$, and hence, by (151) and (201),

$$\begin{aligned} |f(x, n)| &= \left| e(ny) f(h/q, n) - 2\pi i y \int_0^n e(uy) f(h/q, u)\, du \right| \\ &\leqslant (1 + 2\pi)\, n \log^{-3} n, \end{aligned}$$

which proves (152).

3·5. By (152),

$$\left| \int_{E_1} f^3(x, n)\, e(-nx)\, dx \right| \leqslant C_{12} n \log^{-3} n \int_{E_1} |f(x, n)|^2\, dx.$$

Now, by (146),

$$\begin{aligned} \int_{E_1} |f(x, n)|^2\, dx &\leqslant \int_0^1 |f(x, n)|^2\, dx \\ &= \sum_{p<n} \sum_{p'<n} \int_0^1 e\{(p - p')x\}\, dx = \sum_{p<n} 1 = \pi(n), \end{aligned}$$

and, by H.-W., Theorem 7, $\pi(n) \leqslant C_{13} n \log^{-1} n$. Hence

$$\left| \int_{E_1} f^3(x, n)\, e(-nx)\, dx \right| \leqslant C_{14} n^2 \log^{-4} n. \tag{202}$$

3·6. To deal with the case when x *is* near a rational number with a small denominator, we introduce the function $g(x, v)$ defined by

$$g(x,v) = \sum_{2 \leqslant m \leqslant v} \frac{e(mx)}{\log m} \quad (v \geqslant 2) \tag{203}$$

and $g(x, v) = 0$ $(v < 2)$. For any numbers x_1 and x_2,

$$g(x_1+x_2, v) = e(vx_2)\, g(x_1, v) - 2\pi i x_2 \int_0^v e(ux_2)\, g(x_1, u)\, du. \tag{204}$$

This can be proved in the same way as (151).

THEOREM 58. *Let*

$$q \leqslant \log^{15} n, \tag{205}$$

$$|y| \leqslant x_0, \tag{206}$$

and let (155) *hold. Then*

$$\left| f\left(\frac{h}{q}+y, n\right) - \frac{\mu(q)}{\phi(q)} g(y, n) \right| \leqslant n \log^{-69} n.$$

Proof. By (146), $\displaystyle \left| f\left(\frac{h}{q}, v\right) - \sum_{\substack{p \leqslant v \\ p \nmid q}} e\left(\frac{ph}{q}\right) \right| \leqslant \sum_{p \mid q} 1 < q.$

Now, by (42),

$$\sum_{\substack{p \leqslant v \\ p \nmid q}} e\left(\frac{ph}{q}\right) = \sum_{\substack{0 < l \leqslant q \\ (l, q) = 1}} e\left(\frac{lh}{q}\right) \sum_{\substack{p \leqslant v \\ p \equiv l \pmod q}} 1 = \sum_{\substack{0 < l \leqslant q \\ (l, q) = 1}} e\left(\frac{lh}{q}\right) \pi([v]; q, l).$$

Also $\displaystyle \left| \pi([v]; q, l) - \frac{\mathrm{ls}[v]}{\phi(q)} \right| \leqslant C_{15} n \exp\left(-\frac{\sqrt{\log n}}{200}\right) < n \log^{-100} n$

$$(0 \leqslant v \leqslant n,\ (q, l) = 1).$$

This is trivial if $0 \leqslant v \leqslant \sqrt{n}$, and follows from (145), (205) and Theorem 55 (with $u = 16$) if $\sqrt{n} < v \leqslant n$. Finally, by (41) and (203), $\mathrm{ls}[v] = g(0, v)$, and, by (155) and H.-W., Theorem 271,

$$\sum_{\substack{0 < l \leqslant q \\ (l, q) = 1}} e\left(\frac{lh}{q}\right) = \mu(q).$$

Hence, by (205),

$$\left|f\left(\frac{h}{q},v\right)-\frac{\mu(q)}{\phi(q)}g(0,v)\right|<q+\left|\sum_{\substack{p\leqslant v\\ p\nmid q}}e\left(\frac{ph}{q}\right)-\frac{\mu(q)}{\phi(q)}\operatorname{ls}[v]\right|$$

$$=q+\left|\sum_{\substack{0<l\leqslant q\\ (l,q)=1}}e\left(\frac{lh}{q}\right)\left\{\pi([v];q,l)-\frac{\operatorname{ls}[v]}{\phi(q)}\right\}\right|$$

$$\leqslant q+qn\log^{-100}n<2n\log^{-85}n\quad(0\leqslant v\leqslant n).$$

Also, by (151) and (204),

$$f\left(\frac{h}{q}+y,n\right)=e(ny)f\left(\frac{h}{q},n\right)-2\pi iy\int_0^n e(vy)f\left(\frac{h}{q},v\right)dv$$

and $$g(y,n)=e(ny)g(0,n)-2\pi iy\int_0^n e(vy)g(0,v)\,dv.$$

Hence, by (206), (148), and (145),

$$\left|f\left(\frac{h}{q}+y,n\right)-\frac{\mu(q)}{\phi(q)}g(y,n)\right|=\left|e(ny)\left\{f\left(\frac{h}{q},n\right)-\frac{\mu(q)}{\phi(q)}g(0,n)\right\}\right.$$

$$\left.-2\pi iy\int_0^n e(vy)\left\{f\left(\frac{h}{q},v\right)-\frac{\mu(q)}{\phi(q)}g(0,v)\right\}dv\right|$$

$$\leqslant\left|f\left(\frac{h}{q},n\right)-\frac{\mu(q)}{\phi(q)}g(0,n)\right|+2\pi x_0\int_0^n\left|f\left(\frac{h}{q},v\right)-\frac{\mu(q)}{\phi(q)}g(0,v)\right|dv$$

$$\leqslant 2n\log^{-85}n(1+2\pi x_0 n)<14n\log^{-70}n<n\log^{-69}n,$$

which proves the theorem.

3·7. Theorem 59. *Let* (155), (205), *and* (206) *hold. Then*

$$\left|f^3\left(\frac{h}{q}+y,n\right)-\frac{\mu(q)}{\phi^3(q)}g^3(y,n)\right|\leqslant 3n^3\log^{-69}n.$$

This follows from Theorem 58 and the trivial inequalities $|f(x,n)|\leqslant n$ and $|g(y,n)|\leqslant n$ on noting that, if $|z|\leqslant c$ and $|w|\leqslant c$, then $|z^3-w^3|\leqslant 3c^2|z-w|$.

Now, by (150),

$$J(h,q) = e\left(-\frac{nh}{q}\right)\int_{-x_0}^{x_0} f^3\left(\frac{h}{q}+y, n\right) e(-ny)\,dy.$$

Hence, putting $$J_1 = \int_{-x_0}^{x_0} g^3(y,n)\,e(-ny)\,dy, \tag{207}$$

we have, by Theorem 59 and (148),

$$\left|J(h,q) - \frac{\mu(q)}{\phi^3(q)} J_1 e\left(-\frac{nh}{q}\right)\right| \leqslant 6n^2 \log^{-54} n, \tag{208}$$

provided that (155) and (205) hold.

3·8. Let

$$\rho(n) = \sum_{m_1, m_2, m_3} \log^{-1} m_1 \log^{-1} m_2 \log^{-1} m_3 \tag{209}$$

with the conditions of summation $m_1 \geqslant 2$, $m_2 \geqslant 2$, $m_3 \geqslant 2$, and $m_1 + m_2 + m_3 = n$. Then

$$\rho(n) = \int_{-\frac{1}{2}}^{\frac{1}{2}} g^3(y,n)\,e(-ny)\,dy. \tag{210}$$

This follows from (203) in the same way as (147) from (146).

The number of terms on the right-hand side of (209) is $\frac{1}{2}(n-4)(n-5)$, and each term is greater than $\log^{-3} n$ and less than 1. Hence

$$\tfrac{1}{3}n^2 \log^{-3} n < \rho(n) < n^2. \tag{211}$$

Now

$$\left|\sum_{m=2}^{m_1} e(my)\right| \leqslant \frac{1}{|\sin \pi y|} \leqslant \frac{1}{2|y|} \quad (m_1 \geqslant 2,\ 0 < |y| \leqslant \tfrac{1}{2}).$$

Hence, by (203) and Abel's lemma,

$$|g(y,n)| < |y|^{-1} \quad (0 < |y| \leqslant \tfrac{1}{2}),$$

and hence, by (210), (207) and (148),

$$|\rho(n) - J_1| \leqslant 2\int_{x_0}^{\frac{1}{2}} y^{-3}\,dy < x_0^{-2} = n^2 \log^{-30} n.$$

From this and (208) it follows that

$$\left|J(h,q)-\frac{\mu(q)}{\phi^3(q)}\rho(n)e\left(-\frac{nh}{q}\right)\right| \leqslant 6n^2\log^{-24}n+n^2\phi^{-3}(q)\log^{-30}n, \tag{212}$$

again provided that (155) and (205) hold.

3·9. We put (cf. H.-W., § 5·6(2))

$$\sum_{\substack{0<h\leqslant q\\(h,q)=1}} e\left(-\frac{nh}{q}\right)=c_q(n). \tag{213}$$

Then, by (212),

$$\left|\sum_{q\leqslant\log^{15}n}\ \sum_{\substack{0<h\leqslant q\\(h,q)=1}} J(h,q)-\rho(n)\sum_{q\leqslant\log^{15}n}\frac{\mu(q)}{\phi^3(q)}c_q(n)\right|$$
$$\leqslant 6n^2\log^{-24}n+n^2\log^{-30}n\sum_{q\leqslant\log^{15}n}\phi^{-2}(q)<7n^2\log^{-24}n.$$

Hence, by (149) and (202),

$$\left|r(n)-\rho(n)\sum_{q\leqslant\log^{15}n}\frac{\mu(q)}{\phi^3(q)}c_q(n)\right|\leqslant C_{16}n^2\log^{-4}n. \tag{214}$$

Let
$$S(n)=\sum_{q=1}^{\infty}\frac{\mu(q)}{\phi^3(q)}c_q(n). \tag{215}$$

Then, by (213) and H.-W., Theorem 327,

$$\left|S(n)-\sum_{q\leqslant\log^{15}n}\frac{\mu(q)}{\phi^3(q)}c_q(n)\right|\leqslant\sum_{q>\log^{15}n}\phi^{-2}(q)\leqslant C_{17}\log^{-14}n.$$

From this and (214) and (211) it follows that

$$|r(n)-S(n)\rho(n)|\leqslant C_{18}n^2\log^{-4}n. \tag{216}$$

3·10. It remains to deal with $S(n)$. By H.-W., Theorems 60, 67, and 263, $\mu(q)\phi^{-3}(q)\,c_q(n)$ is a multiplicative function of q, and it is trivial that the series in (215) converges absolutely.

Hence, by (215) and Theorem 2,

$$S(n) = \prod_p \left(1 - \frac{c_p(n)}{(p-1)^3}\right).$$

When n is even, this gives $S(n) = 0$, but when n is odd, we obtain

$$S(n) = 2 \prod_{p>2} \left(1 - \frac{c_p(n)}{(p-1)^3}\right) \geqslant 2 \prod_{p>2} \{1 - (p-1)^{-2}\}$$

$$\geqslant 2 \prod_{m=2}^{\infty} (1 - m^{-2}) = 1.$$

From this and (211) and (216) we obtain (144), as we may assume that $C_{11} > e^{3C_{10}}$.

THEOREMS AND FORMULAE FOR REFERENCE

THEOREMS

1. $\zeta(s)$ is regular for $\sigma > 0$, except at $s = 1$, where it has a simple pole with residue 1.

2. If $f(n)$ is multiplicative, and $\sum_{n=1}^{\infty} |f(n)|$ converges, then

$$\sum_{n=1}^{\infty} f(n) = \prod_{p} \sum_{m=0}^{\infty} f(p^m).$$

3. Let $\sigma > 1$. Then

$$\zeta(s) = \prod_{p} (1-p^{-s})^{-1}.$$

5. Let $|z| = 1$. Then

$$\mathbf{R}(3+4z+z^2) \geqslant 0.$$

6. Let $u > 1$. Then

$$\mathbf{R}\left\{3\frac{\zeta'(u)}{\zeta(u)} + 4\frac{\zeta'(u+iv)}{\zeta(u+iv)} + \frac{\zeta'(u+2iv)}{\zeta(u+2iv)}\right\} \leqslant 0$$

and

$$|\zeta^3(u)\,\zeta^4(u+iv)\,\zeta(u+2iv)| \geqslant 1.$$

7. Let $r > 1$, $f(z) = \sum_{n=1}^{\infty} b_n z^n$ $(|z| < r)$, and $\mathbf{R}f(z) \leqslant M$ $(|z| = 1)$. Then

$$|b_n| \leqslant 2M \quad (n = 1, 2, \ldots).$$

9. Let $f(z)$ be regular, and $|f(z)/f(0)| \leqslant e^M$, for $|z| \leqslant 2$; let $0 < a \leqslant 1$, $f(z) \neq 0$ $(|z| \leqslant 1, \mathbf{R}z > 0)$, and let $f(z)$ have a zero of order h at $z = -a$. Then

$$-\mathbf{R}\{f'(0)/f(0)\} \leqslant 2M - h/a.$$

10. Let $f(z)$ be regular, and $|f(z)/f(0)| \leqslant e^M$, for $|z| \leqslant 2$; let $|a| \leqslant 1$, $|b| \leqslant 1$, $a \neq b$, $f(a) = f(b) = 0$, and $f(z) \neq 0$ $(|z| \leqslant 1, \mathbf{R}z > 0)$. Then

$$-\mathbf{R}\{f'(0)/f(0)\} \leqslant 2M + \mathbf{R}(1/a) + \mathbf{R}(1/b).$$

11. $\zeta(s)$ has no zeros in the set of points D given by

$$\sigma > 1 - 1/(4000 \log t^*).$$

12. $\eta'(s)/\eta(s)$ is regular in the set of points D of Theorem 11.

13. Let $0 < r_1 < r_2$, let $g(z)$ be regular for $|z| < r_2$, and let $g(0) = 0$ and $\mathbf{R}g(z) \leqslant M$ $(|z| = r_1)$. Then

$$|g'(z)| \leqslant 2Mr_1(r_1 - |z|)^{-2} \quad (|z| < r_1).$$

14. Let $1 - 1/(10000 \log t^*) \leqslant \sigma < 2$. Then

$$|\eta'(s)/\eta(s)| \leqslant C_2 \log^3 t^*.$$

15. Let $a > 0$, $b > 0$, $x > 0$, and $x \neq 1$. Then

$$\left| \int_{a-ib}^{a+ib} \frac{x^s}{s}\, ds - 2\pi i E(x) \right| \leqslant \frac{2x^a}{b\,|\log x|}.$$

17. Let $m \geqslant 3$ and $a = 1 + 1/\log(m + \frac{1}{2})$. Then

$$\sum_{n=1}^{\infty} \left(\frac{m+\frac{1}{2}}{n}\right)^a \left|\log \frac{m+\frac{1}{2}}{n}\right|^{-1} \log n < 22m(3 + \log m)^2.$$

19.
$$|\pi(m) - \operatorname{ls} m| \leqslant 3C_6 m \exp\left(-\frac{\sqrt{\log m}}{200}\right) \quad (m \geqslant 1).$$

21. Let χ be a character $(\operatorname{mod} k)$. Then

$$\sum_{n=1}^{k} \chi(n) = \phi(k) \text{ or } 0$$

according as χ is or is not the principal character.

22. Let χ_1 be any character $(\operatorname{mod} k)$. Then

$$\sum_{\chi (\operatorname{mod} k)} \chi(n) = \sum_{\chi (\operatorname{mod} k)} \chi_1(n) \chi(n) \quad \text{for any } n.$$

23. Let $m \not\equiv 1$ $(\operatorname{mod} k)$. Then there is a character χ $(\operatorname{mod} k)$ such that $\chi(m) \neq 1$.

35. Let χ be a non-principal character $(\operatorname{mod} k)$. Then $L(s, \chi)$ is regular for $\sigma > 0$, and $|L(s, \chi)| \leqslant k|s|/\sigma$ $(\sigma > 0)$.

36. Let χ be a non-principal character $(\operatorname{mod} k)$. Then

$$|L(1, \chi)| \leqslant 2 + \log k.$$

38. Let $u>1$, and let χ be any real character. Then

$$\mathbf{R}\left\{3\frac{\zeta'(u)}{\zeta(u)}+4\frac{L'(u+iv,\chi)}{L(u+iv,\chi)}+\frac{\zeta'(u+2iv)}{\zeta(u+2iv)}\right\}\leqslant 0$$

and $$|\zeta^3(u)L^4(u+iv,\chi)\zeta(u+2iv)|\geqslant 1.$$

39. Let χ be a non-real character (mod k). Then

$$L(s,\chi)\neq 0 \quad (\sigma>1-1/(4000\log\tau)).$$

40. Let χ be a real non-principal character (mod k). Then

$$L(s,\chi)\neq 0 \quad (\sigma>1-1/(8000\log\tau), t\neq 0).$$

43. Corresponding to any two real characters χ_0 and χ, there is a function $f(n)$, such that $f(n)\geqslant 0$ $(n=1,2,\ldots)$, $f(1)=1$,

and $$\zeta(s)L(s,\chi_0)L(s,\chi)L(s,\chi_0\chi)=\sum_{n=1}^{\infty}f(n)n^{-s} \quad (\sigma>1).$$

44. Let χ be any real non-principal character. Then

$$L(1,\chi)\neq 0.$$

45. Any two L functions formed with equivalent non-principal characters have the same zeros in the half-plane $\sigma>0$.

46. Let $k\geqslant 8$, let χ be a non-principal character (mod k), and let $s>1-1/\log k$. Then

$$|L'(s,\chi)|\leqslant 6\log^2 k.$$

47. Let χ_0 and χ be non-equivalent non-principal real characters (mod k_0) and (mod k) respectively, and let

$$g(s)=\zeta(s)L(s,\chi_0)L(s,\chi)L(s,\chi_0\chi),\ \rho=L(1,\chi_0)L(1,\chi)L(1,\chi_0\chi),$$

and $\frac{9}{10}<a<1$. Then

$$g(a)>\frac{1}{2}-\frac{6\rho}{1-a}(k_0k)^{8(1-a)}.$$

48. Corresponding to any ϵ, there is a positive number A such that, if $k>A$ and χ is a real non-principal character (mod k), then $L(s,\chi)$ has no zeros on the real axis between $1-k^{-\epsilon}$ and 1.

55. Let $m \geqslant 3$, $k \leqslant \log^{u} m$, and $(k, l) = 1$. Then

$$\left|\pi(m; k, l) - \frac{\operatorname{ls} m}{\phi(k)}\right| \leqslant A_8 m \exp\left(-\frac{\sqrt{\log m}}{200}\right).$$

56. Let $n \log^{-3} n < v \leqslant n$, $\log^{15} n < q \leqslant n \log^{-15} n$, and $(h, q) = 1$. Then

$$|f(h/q, v)| \leqslant n \log^{-3} n.$$

FORMULAE

(1) $$\pi(m) = \sum_{n=2}^{m} \frac{1}{\log n} + O(me^{-c\sqrt{\log m}}).$$

(2) $$\psi(m) = m + O(me^{-c\sqrt{\log m}}).$$

(3) $$\psi(m) = \sum_{n=1}^{m} \Lambda(n).$$

(4) $$\zeta(s) = \sum_{n=1}^{\infty} n^{-s} \quad (\sigma > 1).$$

(5) $$\zeta(s) \neq 0 \quad \left(\sigma > 1 - \frac{1}{C_1 \log|t|}, |t| > C_1\right).$$

(9) $$\left|\zeta(s) - \frac{1}{s-1}\right| \leqslant \frac{|s|}{\sigma} \quad (\sigma > 0, s \neq 1).$$

(10) $$\zeta(s) \neq 0 \quad (\sigma > 1).$$

(11) $$\frac{\zeta'(s)}{\zeta(s)} = -\sum_{n=1}^{\infty} \Lambda(n) n^{-s} \quad (\sigma > 1).$$

(14) $$f(z) \neq 0 \quad (|z| \leqslant 1, \mathrm{R}z > 0).$$

(15) $$t^* = \max(|t|, 100).$$

(16) $$\eta(s) = (s-1)\zeta(s) \quad (s \neq 1), \quad \eta(1) = 1.$$

(23) $$|\eta(s)| < \tfrac{4}{3} \quad (1 < u < \tfrac{21}{20}, |s-u| \leqslant \tfrac{1}{4}).$$

(33) $$h \geqslant 4.$$

(35) $$E(x) = \begin{cases} 1 & (x > 1) \\ 0 & (0 < x < 1). \end{cases}$$

(41) $$\operatorname{ls} m = \sum_{n=2}^{m} \frac{1}{\log n} \quad (m \geqslant 2), \ \operatorname{ls} 1 = \operatorname{ls} 0 = 0.$$

(42) $$\pi(m;\ k, l) = \sum_{\substack{p \leqslant m \\ p \equiv l (\mathrm{mod}\ k)}} 1.$$

(45) $$g(n) = \frac{1}{\phi(k)} \sum_{h=1}^{\phi(k)} \overline{\chi}_h(l)\, \chi_h(n).$$

(46) $$\pi(m; \chi) = \sum_{p \leqslant m} \chi(p).$$

(48) $$\pi(m) - \pi(m, \chi_1) \leqslant \phi(k).$$

(48a) $$\pi(m; \chi_h) = O(me^{-c\sqrt{\log m}}) \quad (h = 2, 3, \ldots, \phi(k)).$$

(49) $$L(s, \chi) = \sum_{n=1}^{\infty} \chi(n)\, n^{-s}.$$

(65) $$|\, L(1, \chi)\, | \leqslant k.$$

(66) $$L(s, \chi) = \prod_{p} \{1 - \chi(p)\, p^{-s}\}^{-1} \quad (\sigma > 1).$$

(67) $$L(s, \chi) \neq 0 \quad (\sigma > 1).$$

(68) $$\frac{L'(s, \chi)}{L(s, \chi)} = -\sum_{n=1}^{\infty} \Lambda(n)\, \chi(n)\, n^{-s} \quad (\sigma > 1).$$

(69) $$\tau = \max(\,|\, t\,|, k, 100).$$

(70) $$L(\sigma_0 + it_0, \chi) = 0.$$

(71) $$\sigma_0 > 1 - 1/(4000 \log \tau_0).$$

(72) $$\tau_0 = \max(\,|\, t_0\,|, k, 100).$$

(73) $$\sigma_0 \leqslant 1.$$

(74) $$|\, L(s, \chi)\, | \leqslant \tfrac{3}{2}\tau_0^{\frac{3}{2}} \quad (1 < u < 2, |\, s - u - it_0\,| \leqslant \tfrac{1}{4}).$$

(76) $$u = 1 + 1/(800 \log \tau_0).$$

(77) $$f(z) = \eta^3(u + \tfrac{1}{8}z)\, L^4(u + it_0 + \tfrac{1}{8}z, \chi)\, L(u + 2it_0 + \tfrac{1}{8}z, \chi^2).$$

(84) $$|f(z)/f(0)\,| \leqslant \tau_0^{33/2} \quad (\,|\, z\,| \leqslant 2).$$

(85) $$t_0 \neq 0.$$

(86) $$\sigma_0 > 1 - 1/(8000 \log \tau_0).$$

(88) $$k_0 = \max(k, 100).$$

(110) $$f(n) \geqslant 0 \quad (n = 1, 2, \ldots).$$

(114) $$f(1) = 1.$$

(128) $$b > A_1.$$

(130) $$k \leqslant \log^{3u} b.$$

(144) $$r(n) > 0 \quad (2 \nmid n, n > C_{11}).$$

(145) $$n > C_{11}.$$

(146) $$f(x, v) = \sum_{p \leqslant v} e(px) \quad (v \geqslant 0).$$

(147) $$r(n) = \int_{x_0}^{x_0+1} f^3(x, n)\, e(-nx)\, dx.$$

(148) $$x_0 = n^{-1} \log^{15} n.$$

(149) $$r(n) = \int_{E_1} f^3(x, n)\, e(-nx)\, dx + \sum_{q \leqslant \log^{15} n} \sum_{\substack{0 < h \leqslant q \\ (h, q) = 1}} J(h, q).$$

(150) $$J(h, q) = \int_{h/q - x_0}^{h/q + x_0} f^3(x, n)\, e(-nx)\, dx.$$

(151) $$f(x_1 + x_2, v) = e(vx_2) f(x_1, v) - 2\pi i x_2 \int_0^v e(ux_2) f(x_1, u)\, du.$$

(152) $$|f(x, n)| \leqslant C_{12} n \log^{-3} n \quad (x \in E_1).$$

(154) $$\log^{15} n < q \leqslant n \log^{-15} n.$$

(155) $$(h, q) = 1.$$

(157) $$a_1 = \prod_{p \leqslant \sqrt{n}} p.$$

(159) $$|f(h/q, v) - b_1| \leqslant \sqrt{n}.$$

(160) $$b_1 = b_2 + b_3.$$

(166) $$|b_2| \leqslant 3n \log^{-4} n.$$

(168) $$b_3 = \sum_{k < \log^6 n} b_5(k).$$

(170) $$b_5(k) = b_6 + b_7.$$

$$|b_6| \leqslant n \log^{-100} n. \tag{175}$$

$$b_7 = b_8(v/k) - b_8(v \log^{-5} n) \quad (k < \log^5 n). \tag{177}$$

$$0 < x \leqslant n/k. \tag{178}$$

$$b_8(x) = \sum_{m \leqslant \log n} b_9(m). \tag{179}$$

$$\left| \int_{E_1} f^3(x, n)\, e(-nx)\, dx \right| \leqslant C_{14} n^2 \log^{-4} n. \tag{202}$$

$$g(x, v) = \sum_{2 \leqslant m \leqslant v} \frac{e(mx)}{\log m} \quad (v \geqslant 2). \tag{203}$$

$$q \leqslant \log^{15} n. \tag{205}$$